# Feeding the Ten Billion

## Plants and population growth

At the current rate of increase, the world's population is likely to reach ten billion by the middle of the 21st century. What will be the challenges posed by feeding this population and how can they be addressed? Written to mark the 200th anniversary of the publication of Malthus' seminal '*Essay on the Principle of Population*', this fascinating book looks at the intimate links between population growth and agricultural innovation over the past 10,000 years, providing a series of vignettes which illustrate how the evolution of agriculture has both shaped and been shaped by the course of world population growth. This historical context serves to illuminate our present position and to aid understanding of possible future paths to food security for the planet. A unique and accessible account of the interaction between population and agriculture, this book will be of interest to a wide range of readers, from students and teachers of agriculture through to policy makers, advisers and members of the general public concerned with global population, food supply, agricultural development, environmental degradation and resource depletion.

Lloyd Evans is a distinguished plant physiologist. His early career included a period at the University of Oxford as a Rhodes Scholar, followed by a fellowship at the California Institute of Technology. He then joined the CSIRO Division of Plant Industry in Canberra, becoming its Chief for seven years. He is a Fellow of the Royal Society of London and a former President of the Australian Academy of Science. He is author of a number of well known texts, including *Crop Physiology* and *Crop Evolution, Adaptation and Yield.*

# Feeding the Ten Billion

## Plants and population growth

L.T. EVANS

PUBLISHED BY THE PRESS SYNDICATE OF THE UNIVERSITY OF CAMBRIDGE
The Pitt Building, Trumpington Street, Cambridge CB2 1RP, United Kingdom

CAMBRIDGE UNIVERSITY PRESS
The Edinburgh Building, Cambridge CB2 2RU, UK http://www.cup.cam.ac.uk
40 West 20th Street, New York, NY 10011-4211, USA http://www.cup.org
10 Stamford Road, Oakleigh, Melbourne 3166, Australia



First published 1998

Printed in the United Kingdom at the University Press, Cambridge

Typeset in Utopia 9.5/13pt, in QuarkXpress™ [SE]

*A catalogue record for this book is available from the British Library*

*Library of Congress Cataloguing in Publication data*

Evans, L.T.
Feeding the ten billion : plants and population growth / L.T. Evans.
p. cm.
Includes bibliographical references (p. ) and index.
ISBN 0 521 64081 4. – ISBN 0 521 64685 5 (pbk).
1. Population. 2. Agriculture – Economic aspects.
3. Overpopulation. 4. Food supply. I. Title.
HB871.E92 1998
363.9–dc21 98-26457 CIP

ISBN 0 521 64081 4 hardback
ISBN 0 521 64685 5 paperback

Dedicated to the memory of Thomas Robert Malthus,
two hundred years after the publication of his first
*Essay on the Principle of Population* (1798)

'*Time is the greatest innovator*'

Francis Bacon *Essays* No. 24 (1625)

# *Contents*

# *Preface*

Since long before the publication in 1798 of Robert Malthus' *Essay on the Principle of Population* there has been anxiety that our numbers might soon outstrip our food supply. Even when there were only a quarter of a billion of us on earth, Euripides voiced concern 'to ease the teeming earth of her burden of men'. As our numbers approached one billion, Malthus and his Chinese counterpart, Hung Liang-chi, expressed their doubts as to how many of us could be fed. After two billion arrived, Aldous Huxley, William Vogt and others were pessimistic: we had to tighten our Malthusian belts. In 1951 Le Gros Clark and Pirie asked 'How shall we work the miracle of feeding the four billions?' Yet those billions had accumulated by 1975, better fed than ever in spite of the Paddock brothers' prediction of *Famine–1975!*. More recently, Lester Brown has issued his annual jeremiads in the World Watch Institute's *State of the World* reports as the world population continues to grow.

There have also been optimists, of course, some of whom have estimated that the world could feed more than 50 billion people and who, we trust, will never be proved wrong. Between these vocal extremes there have been well-informed agriculturists, such as Sir John Russell in 1954 and Daniel Hillel in 1991, who have expressed 'tempered' or 'conditional' optimism about the future of the world's food supply. What Malthus achieved with his mathematically simple but emotively powerful contrast between the capacity for increase in our food supply and that in the human population was to ensure that the problems of raising agricultural production continued to gain the attention they needed, until recently.

For a variety of reasons that no longer seems to be so. At two conferences on the world population problem which I attended, food supply was not mentioned as either a constraint or a hazard. The funding of agricultural research in many developing and developed countries has suffered severe cuts in recent years while environmental and social equity issues have moved to centre stage. There are many reasons for this shift, including fashion, donor fatigue and the accumulation of agricultural surpluses. The enormous success of the 'Green Revolution' in the 1960s and 1970s may have lulled governments of both more and less developed countries into believing that further green revolutions can be summoned up as required, especially with the aid of biotechnology. That, at least, is the view that has been put to me by several demographers, economists and public policy makers.

It is easy to see why both laypeople and governments are confused about the capacity of the earth to feed us. In this book, which is not a history of agriculture but a series of vignettes along the way, I try to link the multiplication of our species with some of the advances in the domestication, adaptation, improvement and management of our food crops. The focus is on plants, not animals, and on those crops which either directly or indirectly provide us with most of our food.

By linking population growth with successive advances in agriculture we come to recognize that not only has agricultural evolution made increase in population possible – indeed it has been blamed for it – but also that population growth has driven the development of agriculture, as Ester Boserup supposed. We see some of the great variety of innovations along the way, and of the crucial but often unexpected synergisms between them. We also come to recognize that the path to feeding the ten billion in a sustainable way is still by no means clear.

In writing this book my reach has had to exceed my grasp, in Robert Browning's words. By way of justification, let me quote from Francis Bacon, in *Novum Organon*:

> 'For the history that I require and design, special care is to be taken that it be of wide range and made to the measure of the universe. For the world is not to be narrowed till it will go into the understanding … , but the understanding is to be expanded and opened till it can take in the image of the world.'

*Lloyd Evans*

# *Acknowledgements*

This book has been remoulded, more than they might care to acknowledge, by five friends and colleagues: Professor Dennis Greenland, Dr Margaret Middleton, Dr Tom Neales, Professor Norman Simmonds and my wife Margaret, to all of whom I owe many thanks, including for their help with what Jane Austen referred to as 'lopping and cropping'.

At Cambridge University Press, Alan Crowden encouraged the evolution of this book from an initial gleam in the author's eye, Maria Murphy helped to reshape it and Jo Whelan skilfully removed editorial inconsistencies. I am greatly indebted to the staff of the CSIRO Library at Black Mountain, particularly to Jon Prance, for tracking down many obscure references, and to many colleagues in the CSIRO Division of Plant Industry for illuminating discussions, illustrations and other help, especially to Helen Kaminski for so patiently typing the several drafts.

I thank the Rockefeller Foundation for the award of a month's residency at the Villa Serbelloni in Bellagio, which turned the final editing from a potentially traumatic experience into a delightful one. Insights from many authors have shaped this book. The superscript numbers in the text acknowledge some of these, but to the many others whose works had to be deleted from my too-long list of references as publication loomed, I give thanks and make apology.

I am grateful to the following authors and publishers for permission to use previously published figures: Ester Boserup (Fig. 1b); W.W. Norton & Co. (Fig. 2); John Wiley & Sons (Fig. 4); I.G. Simmons (Fig. 8); the Trustees of the British Museum (Fig. 12); The University of Chicago Press (Fig. 13); The Vatican Library, Rome (Fig. 14); Institut d'Ethnologie,

Musée de l'Homme, Paris (Fig. 15); the Editor of *Nature* (Fig. 16); M.W. Rosegrant for the data used in Fig. 24; FAO, Rome (Fig. 28). I am also grateful to Professor Alec Hope for permission to quote from his poem 'Conversations with Calliope'.

CHAPTER 1

# Introduction: timebomb or treadmill?

Two hundred years ago Robert Malthus had an argument with his father which led directly to the son collecting his thoughts in his *Essay on the Principle of Population*, published in 1798. Malthus (Figure 1a) set out the crux of his argument as follows: 'Population, when unchecked, increases in a geometrical ratio. Subsistence increases only in an arithmetical ratio. A slight acquaintance with numbers will show the immensity of the first power in comparison of the second. By that law of our nature which makes food necessary to the life of man, the effect of these two unequal powers must be kept equal. This implies a strong and constantly operating check on population from the difficulty of subsistence. This difficulty must fall somewhere and must necessarily be severely felt by a large portion of mankind.'[128] Consequently, his father's hopes for an equal society, shared with Godwin and Condorcet, would always founder.

The youthful bravura of his *Essay* caused a considerable stir and soon affected public policies. He became 'the best-abused man of the age'. His insight on the continual trimming of populations by food supply was crucial to both Charles Darwin and A.R. Wallace in formulating their theories of evolution by natural selection, and revisits those concerned by the continuing expansion of the world's human population. In 1845 the Irish potato famine appeared to offer a foretaste of the 'Malthusian spectre' which has given repeated impetus to concerns about the ability of the world's food supply to keep up with population growth.

Since the 1960s, however, the advances made in the yields of the staple food crops and the accumulation of surplus agricultural capacity in several developed countries have led many to conclude that the

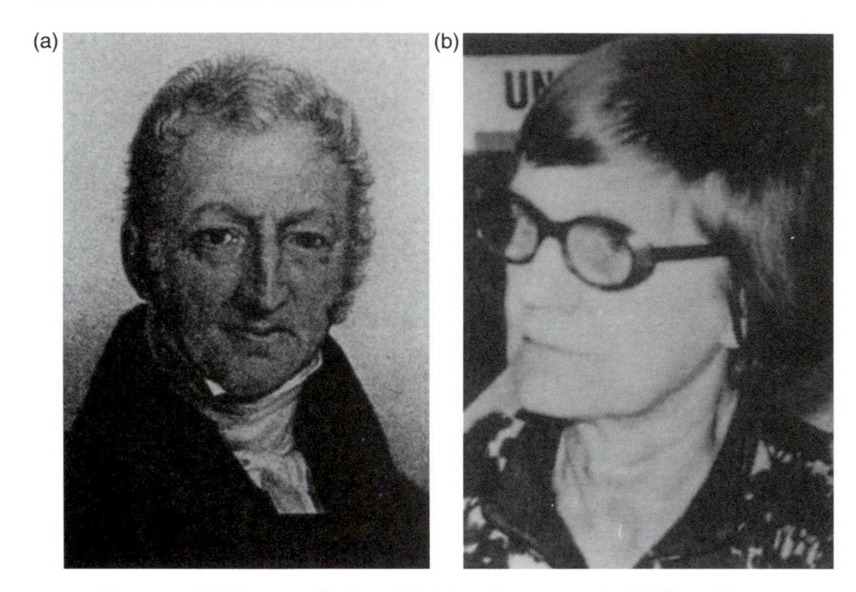


**Figure 1** (a) Thomas Robert Malthus (1766–1834); (b) Ester Boserup (1910– ).

world's food supply problems have, in principle, been solved. For example, in *The World in 2020* McRae[127] writes: 'The old motors of growth – land, capital, natural resources – no longer matter. Land matters little because the rise in agricultural yields has made it possible to produce far more food in the industrial world than it needs … . The capacity of the world to produce enough food to feed a much larger population is not really in doubt …'.

As the real world prices for the staple cereals continue to fall, agricultural science seems to have succeeded too well. Even among concerned biologists there is a widespread impression that, 200 years after it was raised, the Malthusian spectre has been banished and that a rise in cereal prices is all that is needed to elicit the requisite increase in production. Given that a world population of at least ten billion is almost inevitable by the latter half of the 21st century, what are our chances of nearly doubling the world's food supply over the next 50 years or so, and at what cost in terms of long-term sustainability? Some experienced agriculturists think it will be relatively easy, others are pessimistic, but virtually all would be concerned by the likely environmental costs. 'How much land can ten billion people spare for Nature?' is a question recently considered by Paul Waggoner[215].

The quotation from the *Essay* by Malthus given above makes it clear that he regarded the supply of food as the driving variable and population as the dependent one. By contrast, Ester Boserup (Figure 1b) has based her analysis of the relation between agricultural productivity and population growth upon the assumption – which she regards as more realistic and fruitful – 'that the main line of causation is in the opposite direction: population growth is here regarded as the independent variable which in its turn is a major factor determining agricultural developments.'[20]

In fairness to Malthus it should be noted that he also considered this possibility. In the second edition of his *Essay* he wrote[129]: 'agriculture may with more propriety be termed the efficient cause of population than population of agriculture, though they certainly react upon each other, and are mutually necessary to each other's support. This indeed seems to be the hinge on which the subject turns ...'. In the fifth edition of 1817 he added five more paragraphs on this point, first noting 'the tendency to an oscillation or alternation in the increase of population and food in the natural course of their progress'. He then goes on: 'there can be no doubt that in the order of precedence food must take the lead', but adds a consideration of the conditions in which 'the tendency of population to increase, should be one of the most powerful and constant stimulants to agriculture'. We face, it seems, both a Malthusian timebomb and a Boserupian treadmill.

I have quoted Malthus' *Essay* at length because he has, too often, been misrepresented. Increasingly, however, human imagination and understanding, based on research and by no means always driven by population pressure, have become a major accelerator of agricultural progress and, to a much smaller extent, decelerator of population growth. Along with this change in recent years, there has been growing apprehension about the long term health of the agricultural and environmental systems of the world as a whole, evident in the current emphasis on 'sustainability'. Effective husbandry of their resources has long been a major concern of good farmers, but likewise at all stages in the evolution of agriculture farmers have had to balance opportunities and innovations with their costs and risks, often not fully recognized at the time. From the hazards to health in Neolithic villages, through those of salination and siltation in Mesopotamia and of soil loss from the Aegean hills, to the environmental consequences of applications of DDT, fertilizer and herbicide, agriculturists have had to weigh the benefits and costs, both social and environmental, both short and long term, of each innovation.

Another theme which I hope emerges in this book is the continually evolving approach to the age-old and still recurrent problems for agriculture posed by pests, diseases, weeds, fertility maintenance, power supply, crop improvement, etc. Whatever the myths, there has never been a golden age in agriculture when moth and rust did not corrupt, nor will there be a permanent solution to any of these problems. As Geerat Vermeij[213] says of modern species: 'they are not necessarily better able to cope with their enemies than ancient ones were with theirs, but the biological surroundings have in an absolute sense become more challenging.'

Throughout most of this book I shall refer to the evolution, not revolution, of agriculture. Yet agricultural revolutions have often been invoked, from V. Gordon Childe's 'Neolithic revolution' at the dawn of agriculture about 10 thousand (K) years ago to W.S. Gaud's 'Green Revolution' in developing countries in the 1960s. In Western Europe, agricultural revolutions have been claimed for the 6th–9th centuries AD, the 8th–12th centuries, the late 16th century, 1760–1820 in Norfolk, the late 1940s, and more recently for the applications of biotechnology. Besides these there have been Karl Marx's 'agrarian revolutions' involving such aspects as land ownership, farm size and labour supply. Even in the Middle Ages there was more or less continuous change and innovation in agriculture, i.e. evolution. However, at times when several innovations, agrarian as well as agricultural, have interacted synergistically, as they did in Norfolk in the 18th century, the pace of change has accelerated and a revolution can be diagnosed, but even then its geographic spread may be slow and uneven.

My aim in this book is not a history of agriculture, for which there are other sources (e.g. Ernle[57]; Ho[97,98]; Russell[179]; Grigg[77]; Simmons[192]; Hillel[93]; Vasey[212]). Rather, I shall explore the inter-relations between growth of the world population and the evolution of agriculture over the last 10 K years, to illustrate the variety of ways in which agricultural change has occurred and of the stimuli which led to these changes. By doing so I hope to aid public understanding of the challenges facing agriculture in the third century after Malthus.

The growth of the human population is considered here mainly at the global level, although its regional composition has shifted greatly over the years, with first Africa, then successively the Near East, Asia, Europe and, in recent years, the currently developing countries being the dominant region of increase. Estimates of earlier world populations, of which eight time series have been tabulated by Joel Cohen[44], vary to some extent, particularly those for the pre-Christian era, but without changing the overall picture greatly. Three of Cohen's figures, based on the same data but using different scales, are presented in Figure 2. In (a), both scales are arithmetic, which highlights the rapidity of recent growth, but at the expense of all detail before the first half billion of our species were on board Spaceship Earth. In Figure 2(b) the population scale is logarithmic but the time scale remains arithmetic, and we now get a more detailed picture of the changes during the earlier part of the Christian era when population growth was slow and irregular. Besides the recent cycle of rapid population growth since the 17th century, it also indicates what McEvedy & Jones[124] diagnose as a medieval cycle in which the global population almost doubled between 600 and 1300 AD, expanding particularly in Europe.

In Figure 2(c) both scales are now logarithmic, a form of presentation suggested by Edward Deevey[51], which highlights an increase in the number of our hunting and gathering ancestors towards five million (M) about 10 K years ago, followed by a sharp rise to a population approaching a quarter of a billion associated with the shift to agriculture. In this form of presentation, the medieval cycle is barely discernible but the sustained rise since then is striking.

Our aim in this book is to understand how the evolution of agriculture has both shaped and been shaped by the course of world population growth.

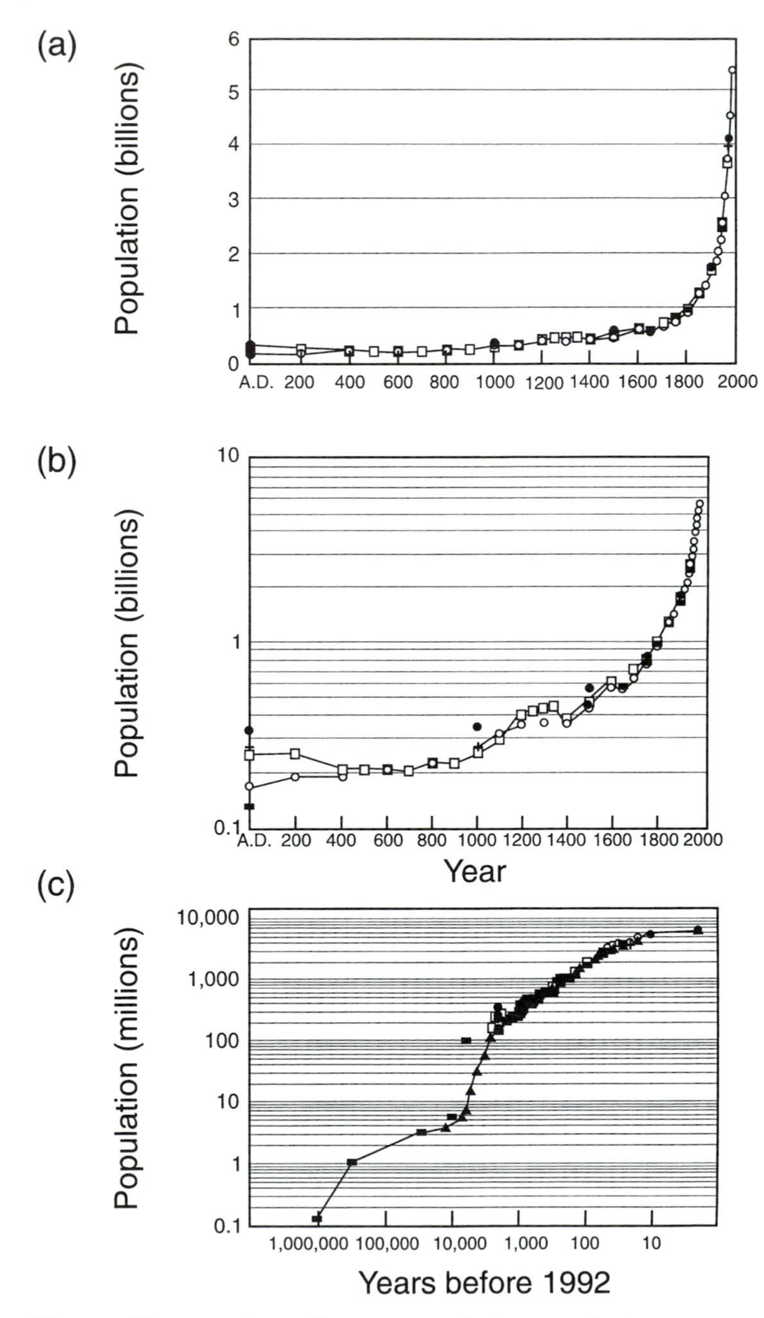


**Figure 2** Three versions of human population growth: (a) arithmetic; (b) semi-logarithmic; (c) both scales logarithmic. (Derived from J.E. Cohen[44].)

# CHAPTER 2

# Reaching five million (TO 8000 BC)

## 2.1 Introduction: the silent millennia

Humankind, i.e. the genus *Homo*, has depended for its subsistence on hunting and gathering for more than 99% of its evolutionary history, yet since that time its population has increased 1000-fold. One demographer has estimated that 12% of the 80 billion humans who have ever lived have done so by hunting and gathering through and after the Pleistocene era. Our information from these silent millennia is slender, and the few surviving groups of hunter–gatherers, such as the Australian aborigines and the !Kung bushmen of the Kalahari desert, have been studied intensively in recent years as 'a window on the Pleistocene', which they manifestly are not.

Given the many uncertainties about human behaviour 10 K years ago, we should not be surprised to find strongly opposed views about population pressure and its role in bringing our total dependence on hunting and gathering to an end. On the one hand there is the view put forward by Childe[40] that the invention of agriculture in the Near East about 10 K years ago made possible not only civilization as we now recognize it but also a rapid increase in the human population. On the other hand there is the Boserupian argument advanced by Mark Cohen in his book *The Food Crisis in Prehistory*[45] for the virtually complete occupation of the world by hunter–gatherers and the build-up in their numbers to the point where, at various places, they were forced by population pressure to turn to agriculture, with considerable reluctance. Deevey[51] and others suggest that there were 5–6 M humans in the world at that time, whereas Cohen suggests 15 M, a level not reached until several thousand years later according to several demographers.

The first primates appeared about 60 M years ago, 5 M years after the demise of the dinosaurs and at about the same time as the wheat, rice and maize genomes began diverging. Hominids and related apes then diverged from the other primates about 12 M years ago. Our upright, bipedal ancestors, *Ardipithecus ramidus* and *Australopithecus anamensis* date back to 4.4 and 4.2 M years respectively. Divergence from the chimpanzees may have occurred only 3.6 M years ago. The more advanced *Australopithecus afarensis*, including 'Lucy', dates back 3.5 M years. 'Handyman' (*Homo habilis*), the first tool maker, appeared about 2.3 M years ago. The oldest tools known, from Ethiopia, date back 2.5 M years. 'Handyman' and/or the later-differentiating *Homo ergaster* began to spread beyond Africa remarkably quickly to reach China almost 2 M years ago and Indonesia (as *Homo erectus*) not much later.

The use of fire associated with human activities, possibly cooking, has been dated as far back as 1.4 M years ago in Africa, and became relatively common 200–300 K years ago when there was an apparent increase in meat eating. DNA sequencing studies now place the origin of modern mankind (*Homo sapiens sapiens*) at about that same time in Africa. Our mitochondrial 'Eve' was deduced to have lived there only 100–200 K years ago and a Y-chromosome 'Adam' about 270 K years ago. But whether modern humans evolved from their local progenitors in parallel ways (the multiregional hypothesis) or replaced them in a later wave out of Africa about 120–150 K years ago (the replacement hypothesis) is still being debated.

In the following years stone and blade tools became predominant, bows and arrows were invented, fire was used not only for cooking but also to manipulate vegetation and facilitate hunting. Grinding stones and digging sticks became quite common by 18 K years ago. At that stage sea level reached its minimum and land area its maximum, connecting islands and continents which are now separate and making migration easier. Then, as the sea level rose again towards its maximum 5 K years ago, much of the habitable land surface was lost, especially in east and south-east Asia where population pressure probably increased.

Our views of the hunting and gathering lifestyle have veered wildly throughout history, from Hesiod's Golden Age and the noble savages of Rousseau to the 'living fossils' of Victorian biologists. Several of the 19th century Australian explorers – such as Eyre, Grey and Giles – were greatly impressed by the food and water-finding skills of the aborigines and by their ability to sustain themselves even under adverse conditions and with relatively little apparent effort. This was confirmed in Arnhem Land in 1948 by McCarthy and McArthur, who found that adult aborigines spent only about four hours per day on average in hunting, collecting and preparing their food, obtaining more than the requisite calories and proteins and a varied diet from enjoyable activities which they did not regard as work even in adverse conditions. When Richard Lee obtained comparable data for the !Kung bushmen of the Kalahari Desert in South Africa, along with evidence of their long life, good health and more stable food supply than that of nearby agricultural populations, the pendulum of opinion about hunting and gathering swung to the other extreme. Marshall Sahlins dubbed them 'the original affluent society'.

> 'No hunter of the Age of Fable
> Had need to buckle in his belt.
> More game than he was ever able
> To take ran wild upon the veldt
> Each night with roast he stocked his table,
> Then procreated on the pelt,
> And that is how, of course, there came,
> At last to be more men than game.'
>
> A.D. Hope, from
> 'Conversations with Calliope'

Today's hunter–gatherers have only the poorest, most difficult environments in which to survive, i.e. those not already claimed for agriculture, grazing or forestry. If they can live so securely in those, they surely lived well in the far more favourable environments available to them until 10 K years ago. The fact that the most closely studied hunting and gathering groups have come from difficult, mostly arid, environments can lead us to underestimate the variety of their lifestyles. Tribes at the higher latitudes lived largely by hunting and fishing, and large animals dominate their cave drawings. At lower latitudes the gathering of plants usually provided most of the dietary calories. Riverine and coastal communities relied on fish and molluscs to a considerable degree throughout the year and developed much higher population densities than communities in arid interiors where seasonal migrations over large areas were the rule, and where up to 250 km$^2$ per person were needed for survival. For these latter, the problems of mobility and of carrying possessions and young children (whether *in-* or *ex-utero*), and the constraints on family size, were far greater than for those in the resource-rich areas.

One characteristic common to all hunter–gatherers is their great knowledge of plants and their life cycles and of animal behaviour. Their survival depended on it. To help things along Australian aborigines sometimes scattered wild millet seeds, replanted yam tops and occasionally 'managed' their wild food plants with fire and even irrigation. In emergencies they might even cultivate yams. Many hunter–gatherers probably had the requisite understanding of plant and animal life cycles to be able to live by agriculture, but preferred the variety of their lifestyle. Cohen[45] therefore concludes that they adopted the agricultural way of life only when driven to it by population pressure.

Such a view overlooks the fact that agriculture would simply not have been a viable alternative for many hunter–gatherers, such as most Australian aborigines and !Kung bushmen. Nor should we forget the likely variety of human preferences among the hunter–gatherers of the world 10 K years ago. While many no doubt enjoyed the freedom of the migratory life, others may have disliked the uncertainties. Small sedentary communities were viable in places where natural resources were abundant before agriculture developed, as we shall see for Abu Hureyra. So the attractions of a more sedentary lifestyle, rather than population pressure, may have led some to agriculture just as others preferred to keep on hunting and gathering.

Finally, there is accumulating archaeological evidence that the hunting/gathering lifestyle was sometimes combined with an element of agriculture, at least in more favourable areas. Such combinations in recent times are known, for example, among the east Brazilian Gé as well as from Austria more than 6 K years ago[79], from Pengtoushan in China where early rice cultivation was combined with the hunting of wild animals more than 8 K years ago and, indeed, from many of the Neolithic villages of the Near East and India. Fire has been used for the control of vegetation for at least 35 K years. It is therefore likely that, at least in the higher rainfall forested areas, plantings of various root crops, bananas, sugar cane, etc. following slash and burn clearing could have been combined with a predominantly hunting, fishing and gathering subsistence. If so, the emergence of agriculture from the hunter–gatherer lifestyle may have been much earlier, more widespread and more gradual than implied by V. Gordon Childe's term, the Neolithic Revolution.

## 2.2 Australian aborigines

Hunting and gathering ancestors of the aborigines entered Australia from south-east Asia over 55 K years ago, when the sea level was more than 100 metres lower than at present. In the monsoonal north they could have found wild yams, wild rice and other food plants familiar to them, and they soon spread down the coastline. About 20 K years ago they began to occupy the drier, more difficult, quintessentially Australian interior of the continent. By the time of European occupation many linguistic and territorial units occupied and 'used' the whole continent. It was by no means *terra nullius*.

No doubt their techniques and customs of gathering and preparing plant foods have changed over the silent millennia. Nevertheless, the recorded observations of many early explorers combined with those of anthropologists in more recent times, and with the still-slender archaeological records, give us our most authentic and comprehensive picture of the hunting and gathering lifestyle.

Of course, the lifestyles vary from one environment to another. On the north coast of Arnhem land, the staple plant foods of the Gidjingali are the roots or rhizomes of several species of water lily, yams and the spike rush, together with the nuts of a cycad. These are supplemented in due season with many other gatherings, such as pandanus nuts, *Livistona* palm pith, arrowroot tubers, etc. Rhys Jones and Betty Meehan estimate that when the total time for gathering and preparing is taken into account, the long yams, pandanus nuts and cycad 'bread' all yielded about 2000 kilocalories per hour, i.e. about one and a half hours of gathering and preparing would meet the daily requirement for energy, confirming the earlier studies by McCarthy and McArthur. They also estimate that the gathering of plant foods by women contributed about half of the energy intake, the other half coming from hunting.

Another favourable site for gathering was described by the explorer Sir George Grey in 1841. At several places in south-west Australia he encountered large 'fields' (i.e. several kilometres wide and long) of fertile soils covered with *Dioscorea* yam plants. These supported a dense and largely sedentary population with 'superior' huts, roads and wells. As in Arnhem land, the yam tops were replanted, but the only digging was to extract the rest of the yam tubers from the ground. Charcoal from several similar yam diggings in Western Australia has been found to be almost 40 K years old.

Even in the drier areas of central Australia *Dioscorea* yams often loom large in the diet, although many other species, including grasses, legumes and fruits are gathered in season. Grass seeds can be collected fairly quickly (e.g. at 2 kg per hour in one study) but their cleaning, winnowing and especially grinding take much longer, e.g. five hours per kilogram. *Acacia* seeds were easier to collect but took even longer to process. Cane estimates that even to provide only 30% of the diet for a family of five would require 10–15 hours of work on grasses each day by the women. This is hardly Sahlins' 'original affluent society', and it is no wonder that grass seeds usually comprised only a small part of the diet.

The grass in this case was a small-seeded millet, and the larger-grained cereals of the Fertile Crescent are likely to have required far less time for cleaning and grinding, although the human bones of Tell Abu Hureyra suggest it was still a heavy task. Despite the effort required in their preparation, grass seeds were an important component of diets in the drier regions for two reasons. They could be stored away for the hungry season, unlike yams and meat, and they could be turned into readily-carried 'breads' for long treks. Small wonder then that in 1886 the explorer A.C. Gregory saw 'fields of 1000 acres' of *Panicum* grass being cut by the natives with stone knives and the seeds swept up, winnowed and ground on stone slabs.

Despite the scale, all this still falls short of agriculture in two respects: the soil was not cultivated, although it was dug for yam harvests, and the plants were not domesticated, i.e. not genetically modified by selection. Yet many of the plants used by the Australian aborigines are wild relatives of crops domesticated elsewhere, such as sorghum, millet, rice, soybean, cotton, yams, sweet potato and tobacco.

Nevertheless, the Australian aborigines came close to agriculture in many other respects. They used and depended on a great variety of plants, up to 30% of the 300 species in their territory in the case of the Alyawara. They distinguished them and knew which were sweet, bitter or poisonous. They had an accurate knowledge of when and where plants would be flowering, fruiting or ready for harvest, as the explorer Edward Palmer recorded in 1883. They had been grinding seeds for many years, and storing them against seasonal shortages. They also managed their 'crop plants'. The tops of yam tubers were replanted, and millet seed was sometimes thrown around after harvest. Drainage was used to control the water level in swamps in southern Australia, and fire was used to control the flowering time of cycads and to promote the growth of yams. As Sir George Grey wrote in 1841: 'the natives must be admitted to bestow a sort of cultivation upon this root ...' They also practised occasional irrigation, but were reluctant weeders.

At least in the north they presumably knew about agriculture from visiting Macassans, and they sometimes acknowledged individual ownership, e.g. of trees. In good yam country they even had more or less permanent settlements. Why then did they not begin agriculture? Possibly the pressure of population on resources may not have built up sufficiently by the time of European occupation. Also, as is apparent still today, many aborigines enjoy the mobile way of life, adjusted to the seasons and to local resources. Moreover, for some groups at least, their complex and integrated beliefs and practices might have precluded the extensive cultivation of land. Chief Seattle long ago expressed the offence to some American Indian sensibilities of ploughing the land.

## 2.3 The !Kung San of Dobe

In his book *The Lost World of the Kalahari* Laurens van der Post referred to the Bushmen as a 'pure remnant of the unique and almost vanished First People of my native land'. As such these Koi San bushmen, who can now practise their ancient lifestyle in only a few arid and isolated areas of Botswana, Angola and Namibia, call themselves the *Zhun/twasi*, 'the real people'.

The five hundred or so !Kung San of the Dobe area were chosen for intensive study in the 1960s by Richard Lee and Irven De Vore partly on the grounds that they were closer than the Australian aborigines 'to the actual floral and faunal environment occupied by early man'. Rather to the surprise of Lee and De Vore, the harsh Kalahari desert was found to provide the !Kung with an adequate and well-balanced diet. Most of the gathering of plant foods from 105 species – including nuts, beans, bulbs, roots, leafy greens, tree resin, berries and fruits – was done by the women and children, in two or three days' work each week. These gatherings provided 60–80% of the diet by weight. Hunting, mainly by the men with spears and arrows poisoned with an extract of beetle larvae, provided the more cherished foods. The !Kung San eat as much vegetable food as they need and as much meat as they can get.

Although average body weight fell in the lean season, hunting and gathering sufficed even in a prolonged drought. Indeed, nearby Herero pastoralists and farmers were forced to join the !Kung in these activities in order to survive. The backbone of their diet was the mongongo tree which produces such an abundance of tasty fruits and nutritious nuts that many rot on the ground in spite of providing more than half of the vegetable diet. When Lee asked a !Kung bushman early in the 1960s why they didn't take up agriculture, he replied: 'Why should we plant when there are so many mongongo nuts in the world?' It was not to last, however. At the end of the project in 1972 Lee estimated that less than 5% of the !Kung San simply hunted and gathered for a living. 'Nisa' told Marjorie Shostak that their traditional way of life was becoming impossible because their lands were being grazed and their wells used for stock. She and her husband wanted to buy goats and plant crops, with little of the reluctance emphasized by Cohen at the prospect of an agricultural way of life.

Even if grazing stock had not intruded on their traditional lands, would the supply of mongongo nuts have remained sufficient without the continuous departure of !Kung people for more modern lifestyles? Lee[115] thought their population could be maintained in equilibrium with their environment, but others are less sure.

Moreover, it now seems that the recent enmeshing of the !Kung with pastoralism and agriculture is not for the first time. There is archaeological evidence of pastoralism, and even of agriculture, in parts of the Kalahari. Pre-Iron Age pastoralism with domesticated cattle, and ceramics, extends back to before 300 AD, and there is also evidence of Iron Age economies and trade in the eastern Kalahari from the 7th to the 11th centuries AD, involving domesticated animals and the growing of sorghum, millet, melons and cowpeas at several localities[52].

The !Kung have therefore had several opportunities to look through this 'window to the Pleistocene' in the other direction and over the years some have preferred to live by hunting and gathering, but many have chosen to live by pastoralism or agriculture or mining[225]. Even among those who chose to hunt and gather, however, the advantages of occasional access to milk for their children, machine-sown clothing, store-bought implements, familiarity with trucks and guns, and occasional wages were not foresworn. Unlike their counterparts in the film *The Gods Must be Crazy*, they know a Coca-Cola bottle when they see one.

As Marjorie Shostak's account of the life of one !Kung woman, *Nisa*, shows so well, the attractions of hunting and gathering could be abandoned and resumed periodically. The reluctance of the !Kung to take up agriculture, so central to Cohen's argument that only population pressure drove hunter–gatherers to it, may reflect their considered judgement that, until recently, agriculture offered little advantage in their environment and that, in any case, they could continue to have a bet each way.

## 2.4 Tell Abu Hureyra

Whatever you might think of irrigation dams, their building is a great instigator of archaeological rescues and investigations, as at Abu Simbel before the Aswan dam was built. In 1972 when the Euphrates river was about to be impounded by a dam at Tabqa in northern Syria, funds were mobilized for the excavation of Tell Abu Hureyra before it was flooded. In the event, the tell was finally submerged a few weeks after digging ceased in 1974. A tell is a mound of debris accumulated during millennia of human occupation, and Abu Hureyra was one of the largest in the Near East, covering almost 12 hectares and rising up to 8 metres high[141].

It turned out that Abu Hureyra had been occupied twice, the first time before and the second time after the Neolithic revolution, so it is one of the few sites where pre- and post-agricultural lifestyles can be compared. The first settlement, at the northern end of the mound, began about 11 K years ago and lasted a millennium, during which the climate was becoming warmer and wetter. Abu Hureyra overlooked the edge of the Euphrates flood plain, lying between steppe to the south and Mediterranean forest to the west. Both kinds of vegetation were rich in plant and animal species suitable for gathering and hunting. Conditions were therefore favourable for a more or less permanent settlement there even before agriculture developed.

The excavations indicated that the first settlement was a group of pit dwellings dug into the ground but with walls and roof (probably made of rushes) supported by wooden poles. Within these dwellings were found an abundance of grinding stones, pestles and flint blades, as well as the charred remains of a remarkable variety of plants and animals. Seeds of at least 157 different plants had contributed to the diet, most notably wild einkorn wheat, barley and (most unusually) rye among the cereal progenitors, and lentils, bitter vetch, hackberry fruits and turpentine nuts[94]. There were also abundant remains of gazelles, goats and sheep, as well as of fish and shellfish from the Euphrates.

But was it just a seasonal camp or a permanent settlement? Gordon Hillman, who has done much of the work on the plant remains, examined the seasonal availability of the various seeds and concluded that the inhabitants must have been at Abu Hureyra at least from March to November and probably for the whole year. Their migration elsewhere for the winter seems unlikely. Because wild einkorn wheat no longer grows near Abu Hureyra, the possibility was raised that the three cereals had been cultivated there, even though their remains still resembled those of their wild relatives. In other words, perhaps they represented the very first step towards domestication. Hillman and his colleagues examined this possibility in several ways and concluded that no, they must have been harvested from wild stands that existed nearby at the time[96].

Why the tell was abandoned for several hundred years is not clear, but it was reoccupied about 9 K years ago[142]. By that time agriculture was widely practised in the Levant, and was undoubtedly the basis for the large community on the tell, the population of which is estimated to have reached several thousands at its peak 8.5 K years ago. It then declined as the climate became drier, the forest retreated and the steppe deteriorated. The tell was finally abandoned about 7 K years ago, not long after pottery made its first appearance there.

In the second occupation the people lived in larger, rectangular houses which were close together, of several rooms, with plastered mudbrick walls which were often whitewashed and even decorated. Such a large community no doubt required some communal organization, and probably had contacts and trade with other sites and even regions. This is suggested by the finding of obsidian from Turkey, turquoise from Sinai and cowrie shells from the Mediterranean.

Although hunting and gathering continued, the staple foods were cultivated einkorn and also now emmer wheat, barley, lentils and chickpeas (among the oldest known). Gazelles were initially abundant in the animal remains, along with sheep and goats, but the gazelles suddenly declined about 9 K years ago, and sheep, goats, cattle and pigs, probably all domesticated, became predominant. The plant remains suggest that the steppe then began to deteriorate, presumably under the combined effects of increasing aridity and over-grazing, and the inhabitants of the tell eventually left for more favourable environments.

Besides the profusion of animal bones on the tell, there were also those of 162 people. On examination these suggested to Theya Molleson[140] that, in general, the people were healthy. There are signs of excessive strain in the vertebrae and neck, especially of the young, presumably from carrying heavy loads, often on their heads. Collapsed vertebrae in the lower back and arthritic big toes, especially among the women and young girls, were deduced by Molleson not to indicate Neolithic ballerinas but to have been caused by long hours on their knees grinding grain. In spite of these efforts, there were many fractured and severely worn teeth. Marks on teeth from gripping reeds while weaving were also apparent in one part of the settlement, suggesting a degree of craft specialization.

Thanks to the Tabqa dam, therefore, we have one of the best examples of the shift from hunting and gathering to agriculture at one site, and of year-round settlement preceding agriculture. Abu Hureyra also illustrates the power of environmental change, in this case from conditions so favourable that sedentary life was possible without agriculture to conditions so unfavourable that it could not continue even with agriculture. The bones and stones tell an interesting story, as do the carbonized remains of plants for those who can read them.

# CHAPTER 3

# Towards fifty million (8000 BC–2000 BC)

## 3.1 Introduction: from foraging to farming

Adam Smith thought mankind passed through an age of hunters and then an age of shepherds before the ages of agriculture and of commerce. However, this no longer seems likely, at least in the Near East where the domestication of crops accompanied that of animals. Nevertheless our hunting and gathering ancestors probably became farmers through a series of steps which should be described if we are to be clear about what we mean by agriculture.

We have already seen that Australian aborigines, like hunter–gatherers elsewhere, sometimes aided the multiplication of the plants they harvested, e.g. by replanting yam tops or scattering millet seed. They even 'managed' these plants occasionally with fire or primitive irrigation. So it was no great step to *cultivation*, which can be defined as the habit of deliberately growing useful plants. Those first cultivated deliberately may have been for ceremonial, medicinal or flavouring purposes, or for hunting (e.g. poisons) or weaving, rather than for food. But undoubtedly some of the seeds gathered for food or receptacles (e.g. gourds) and the remnants of tubers and roots would have given rise to plants in the more fertile and disturbed soils and refuse heaps of early settlements.

Most dictionaries add the tilling and improvement of soil as elements of 'cultivation' and these activities may also have occurred sometimes in the settlements of hunter–gatherers. The digging sticks needed to gather wild yams could also be used to plant surplus tubers near huts. Preferred varieties, such as the easier-to-dig shallow-rooted yams, might have been selected, and occasional tilling and even weeding practised.

In forested areas where fish or game were plentiful and communities more sedentary, the step towards partial dependence on shifting cultivation for starchy foods would certainly not have been revolutionary, nor incompatible with some continuing dependence on hunting and gathering. Land preparation for planting in such areas required only axe, fire and digging stick. Although rice and maize are now common crops in contemporary shifting cultivation systems, the root crops which may have predominated in early ones would have been too low in protein to provide an adequate diet on their own.

Although agriculture (*agri-cultura*) simply meant field cultivation, in a fuller sense it also involved the establishment of an artificial ecosystem to yield an adequate and permanent staple food supply. Agriculture is a system of food, feed and fibre production involving a variety of domesticated (i.e. genetically modified) plants and animals together with a sustained input of human effort for cultivating the soil and for tending and harvesting the crops. An agricultural society is one which has become dependent on cultivated foods and which commits its resources to the maintenance of its agricultural systems. It was in this sense that Childe viewed his Neolithic revolution in the Near East.

What were the forces that drew humankind into dependence on agriculture? In broad terms four should be considered. The first derives from the evolutionary perspective presented above. Human nature being what it is, the shift from hunting and gathering through a series of steps towards agriculture was highly probable in favourable environments and particularly where wild plants, especially grasses, suitable for domestication occurred in abundant stands. In the Near East and also in Mexico the coincidence between the distribution of the earliest agricultural settlements and that of dense stands of the wild cereal progenitors is striking indeed.

As long ago as 1882 the Swiss botanist Alphonse de Candolle concluded from his survey of the origins of cultivated plants that they had given rise to agriculture in three separate regions, China, south-west Asia and tropical America. In 1926 Nicolai Vavilov, the Russian plant geographer, identified eight independent centres of origin of cultivated plants (Figure 3). Vavilov assumed that the centre of greatest genetic diversity within a crop was also its centre of origin, but this was found not to be so. For example, the genetic diversity of barley is greatest in

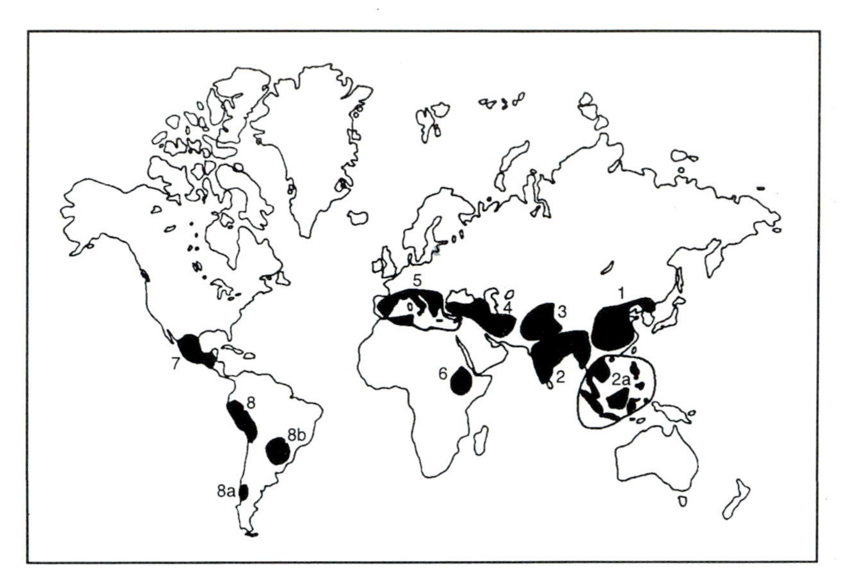


Figure 3 The eight centres of origin of crop plants as proposed by N.I. Vavilov.

Ethiopia yet its origins clearly lie in the Near East, where it is not particularly diverse but where its wild progenitor occurs in abundance. Moreover, while some of Vavilov's centres of origin remain well defined – namely the three which de Candolle had identified – others, such as Africa, are so diffuse that Harlan[83] has called them 'non-centres'.

Nevertheless, the local availability of plants and animals suitable for domestication was clearly a major determinant of the shift towards agriculture, in what Rindos[169] regards as a co-evolutionary process between plants and people, not necessarily in response to stress by either environment or population pressure. He emphasizes the operation of unconscious selection in this process but it seems likely that deliberate selection, whether for grain size in wheat or tuber size or shape in yams, was also important.

A second force driving people to agriculture could have been a change in climate. This was the cause favoured by Childe for what he called the Neolithic Revolution in the Near East. He supposed a widespread change to significantly drier conditions about 10 K years ago, and such changes did occur in some localities, as we have seen for Abu Hureyra. They may also have played a role in Central America and in Africa, but not universally where and when the process of domestication was beginning. Towards the end of the Ice Ages there was a foretaste of the warming to come for about a millennium 15 K years ago. The temperature then fell again for another two millennia before abruptly rising 7°C about 11.5 K years ago and a further 3–4°C over the next 2 K years. Since then we have been living in unusually stable times. For example, the 'little Ice Age' from 1350–1850 AD was only 0.5°C cooler than preceding years, whereas the present global warming may raise the average temperature 1–2°C. Besides the warmer and drier conditions 10 K years ago there were associated changes which may also have played a role, such as the rise in atmospheric $CO_2$ level[180], an increase in seasonality and the possibly greater prominence of annual grasses.

A third driving force could have been the desire of many people for a more sedentary lifestyle than hunting and gathering usually allows. While many hunter–gatherers savour the freedoms and challenges of mobility, others may have preferred the greater security, the opportunity to have more children than is manageable on the move, the less demanding old age, or the possibility of accumulating possessions, owning land, specializing in crafts, laying by for future adversity (e.g. in grain storage) and enjoying a greater variety of human contacts. Sedentary communities existed before agriculture, as at Abu Hureyra and other places where food resources were plentiful, but they could be supported more securely and in a greater range of environments once agriculture developed.

The fourth driving force could have been population pressure, as suggested by Mark Cohen[45]. Much of the momentum for this view came from the change in how the lifestyle of hunter–gatherers was viewed. When they were seen as savages eking out a meagre living from their environment and without time for cultural activities, the invention of agriculture was viewed as an overwhelming attraction to be seized on enthusiastically. But after the lifestyle of the !Kung had been painted as so leisurely, secure and varied, the life of the early farmer was viewed as being so dull and hard-working that it would be adopted only under severe pressure, presumably population pressure.

The demographic evidence for such pressure to have been widespread 10 K years ago is minimal. Given the limitations on family size among mobile hunter–gatherers, e.g. by the prolonged duration of breast feeding (a powerful contraceptive practice), the increase in population must have been slow except in the more sedentary bands. After agriculture was developed, rapid increases in population occurred mainly in those areas where agriculture was adopted, like China, India and the Near East, but not for example in Japan, Korea and Siberia where the people remained as hunter–gatherers[124].

Moreover, we should not assume that early agriculture resulted in pronounced increases in grain supplies. Measurements of the yield of grain harvested from stands of wild wheat in the Near East, of wild rice in India and of teosinte in Mexico have all been in the range of 0.5–1 tonne per hectare of clean grain. Early crop yields are unlikely to have been much greater than this, although better grain retention after domestication should have resulted in more efficient harvesting. At least initially, therefore, the shift to agriculture is unlikely to have greatly increased food supply, but rather to have concentrated it nearer the villages and made it more secure. Childe regarded the possibility of surplus accumulation – whether of stored grain or of meat 'on-the-hoof' – as a crucial element of the Neolithic revolution, both as reserves and as a basis for rudimentary trade. Comparisons of human remains from before and after the transition to agriculture in several regions suggest that although there was a decline in stature and health, mortality decreased among children and the old but birth rates did not increase for some time.

Thus, there is little evidence that the shift to agriculture was driven by population pressure, but rather that it opened the way, later, to the Malthusian scenario. By the end of the agricultural transition on a world-wide scale, about 4 K years ago, a great variety of farming systems had evolved. The human population had increased about ten-fold on a world scale and far more than that in the early agricultural regions. Civilizations developed, competing cities and states made war, social inequalities were elaborated, castes and crafts flourished. The commitment to dependence on agriculture was no longer reversible and, as Alphonse de Candolle pointed out in 1882, humankind had already domesticated virtually all the staple crops on which it was to rely.

## 3.2 Shifting cultivation

The headline in the *New Scientist* (12/12/1992) claimed 'Pacific islanders were world's first farmers'. The slender basis for this claim was the finding of many flake tools, dating back to 28 K years ago, in Kilu Cave on Buka Island in the Solomons, on which Australian archaeologists identified starch grains and raphide crystals like those found in two species of taro, tuber crops of south-east Asia. This would place the origins of agriculture in tropical forests rather than temperate savannahs, with root and tuber rather than seed crops, in south-east Asia rather than the Near East, and by evolution rather than by revolution, as Carl Sauer[184] and others have long argued.

However, the chances of finding carbonized cereal and legume grains in dry regions from settlements 10 K years old are far greater than those of identifying the remains of even older root and tuber crops in wetter sites, many of which may have been submerged by the post-Pleistocene rise in sea level. Although the archaeological evidence is slight, shifting cultivation is almost certainly more than 10 K years old.

Fire has been used extensively by hunter–gatherers to manage vegetation for at least 35 K years. Some anthropologists have doubted whether Stone Age people would be able to fell areas of virgin forest. However, in 1952, several archaeologists cleared a hectare of Danish oak forest using Stone Age axes and, after burning off, grew a luxuriant cereal crop[105]. During subsequent visits to New Guinea, Axel Steensberg[201] recorded the traditional skills of shifting cultivators in the use of stone axes and adzes, bamboo knives and wooden digging sticks in clearing and preparing land for their crops. One of the most labour-saving and spectacular of these skills is the domino-like windrow felling of trees down a slope, a technique used by shifting cultivators throughout the world.

Stone axes at least 26 K years old have been found in New Guinea, and evidence of forest burning there goes back at least 30 K years. Thus shifting cultivation could have been practised there long before the Neolithic 'revolution' in the Near East. Quite an array of crops was domesticated in New Guinea, including sugar cane, many vegetables and fruits, sago, the swamp taro, several yams and the *fe'i* banana. There is evidence for the cultivation of wetland crops in the Kuk area of highland New Guinea about 9 K years ago.

Given the ubiquity and variety of shifting cultivation still practised in the tropics, its techniques are likely to have evolved early and independently in many regions. Although it was also practised in temperate zones, shifting cultivation today is most prominent in the tropics of Asia, Latin America and Africa (Figure 4), particularly in those zones with seasonally dry periods of 3–6 months duration. At the wetter end of this range root and tuber crops such as yams, sweet potatoes and

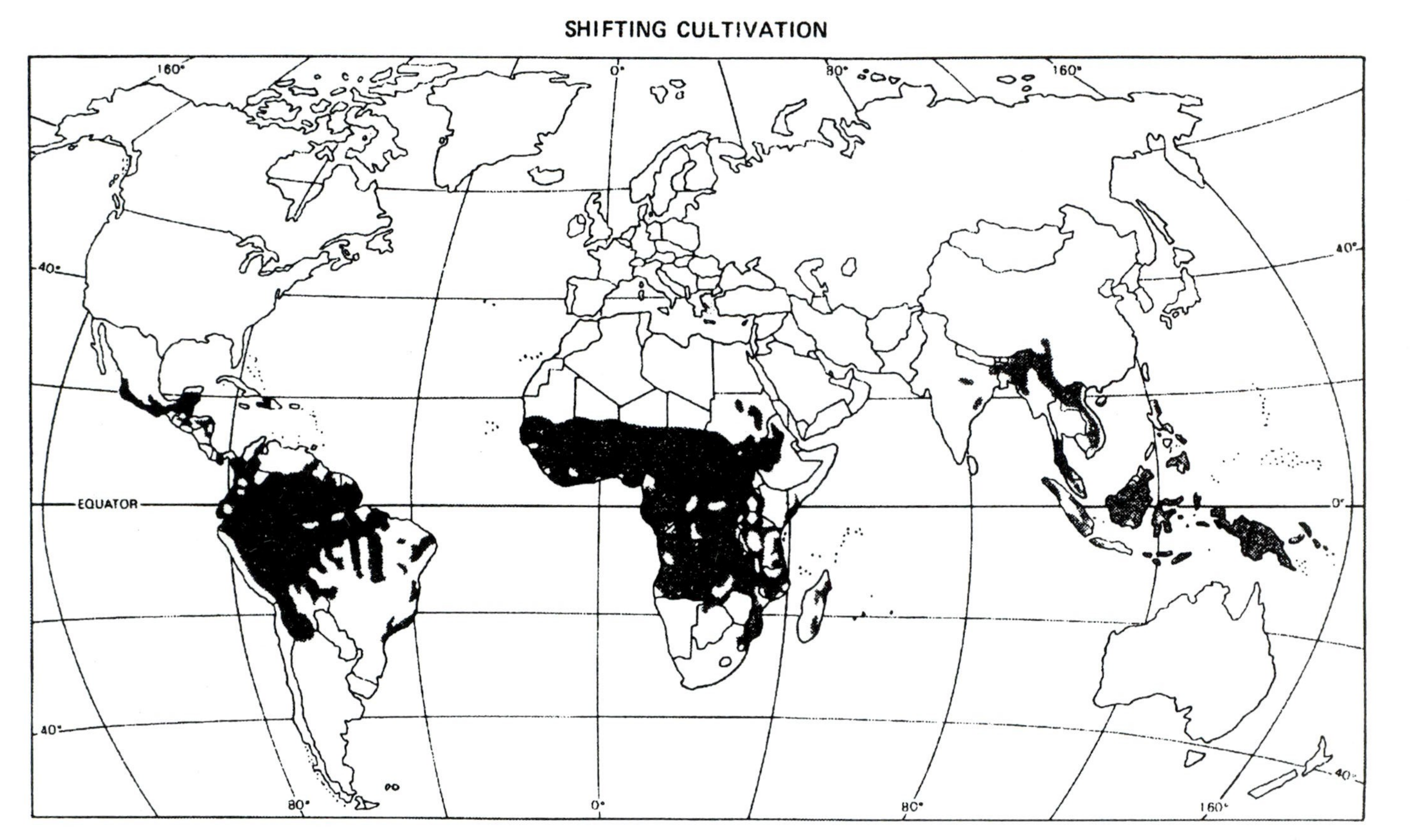


**Figure 4** Areas with shifting cultivation (after J.H. Butler[29]).

cassava (manioc) predominate, whereas cereals such as upland rice, maize and sorghum are more prominent in the drier regions.

The distribution of the many domesticated species of *Dioscorea* yams in America, Africa and Asia corresponds closely with that of tropical areas with relatively short dry seasons. In these regions, yams and many other root and tuber crop plants such as manioc and taro are likely to have played an important role in the evolution of shifting cultivation systems, along with perennial crops such as bananas, sugar cane and sago. Such plants have low protein and mineral contents in their storage organs, which adapts them to soils low in nitrogen and other nutrients, to more prolonged cropping periods and to less expert burns[87]. As such they would have been more suited than the later-domesticated cereals to the early forms of shifting cultivation, but would have provided an adequate diet only when combined with fishing, hunting or the management of grazing animals.

In most shifting cultivation systems soil fertility and pH are highest for the first crop after the burn due to the mineral nutrients previously accumulated in the forest biomass, but then decline[147]. Crops with lower soil fertility requirements, such as manioc, may be planted in subsequent seasons, along with legumes, ground cover plants, fruit-bearing shrubs, bamboos and useful trees which lead back to the multi-tiered, protective and restorative forest and bush fallows. These may also be important managed components of the overall cropping system, as was the ramon nut tree in the Mayan rain forests.

Shifting cultivation often gets a bad press. In 1957 FAO labelled it as the most serious land use problem in the tropics. Blamed for desertification in the 1970s, for deforestation in the 1980s and for non-sustainability in the 1990s, it still supports about 300 M people, i.e. more than the total world population in AD 1, clearing 20–60 M ha per year. Traditional shifting cultivation can be ecologically congruent while meeting a variety of human needs[147]. However, as population pressure increases, the forest fallows of 20–25 years duration give way to less restorative bush fallows of 6–10 years. Further shortening of the fallow period leads to the breakdown of what otherwise can be a productive and sustainable agricultural system. At Yurimaguas in the Amazon basin, for example, the minimum effective period of bush fallow is 12 years, whereas population pressure has reduced it to only 4 years[182], which is better referred to as 'slash and burn' and is a more valid focus for environmentalist ire.

In 1966 the average density for populations reliant on shifting cultivation was 0.06 persons per hectare, to be compared with an average of 3 persons per hectare of arable land then for the world as a whole. Although increasing population pressure has led from long forest fallows to progressively shorter and less effective bush fallows as Boserup[20] recognized, it did not lead directly to annual cropping in the tropics. Rather, that has come from the adaptation of agricultural systems which developed independently in more temperate regions, mostly away from the forests.

## 3.3 The Neolithic Revolution

In the Near East about 10 K years ago several shifts in lifestyle and dependence took place more or less simultaneously (i.e. over a millennium or more!) which eventually had such profound consequences for human civilization that the Australian archaeologist V. Gordon Childe called them 'the Neolithic Revolution'.

Looking back through those ten millennia we can be so awed by the changes that flowed from humanity's shift to dependence on agriculture that we search for major causes. At the time, however, the transition to domesticated crops and animals may have been seen simply as a supplement to hunting and gathering, and a not particularly welcome one at that given the additional labour involved in cultivation, the need to stay close to the stored grain, and the effort required to prepare these foods.

One striking feature of the Near East is the close coincidence between the present-day distribution of the wild progenitors of wheat and barley and the location of the early farming villages[86] across the 'Fertile Crescent', as shown in Figure 5. Some of the pre-agricultural settlements were located in or very close to dense stands of wild wheat or barley whereas many of the neolithic agricultural villages are near or just beyond the margins of present-day occurrence of the wild progenitors. Of course, small changes in climate, intensive gathering by denser populations or heavy grazing by goats may have made present distributions more restricted than those in the past. Wild einkorn wheat no longer grows near Abu Hureyra, wild emmer wheat no longer at Cayonu, wild barley no longer at Hacilar. Villagers living close to abundant wild stands of wheat and barley are less likely to want to cultivate

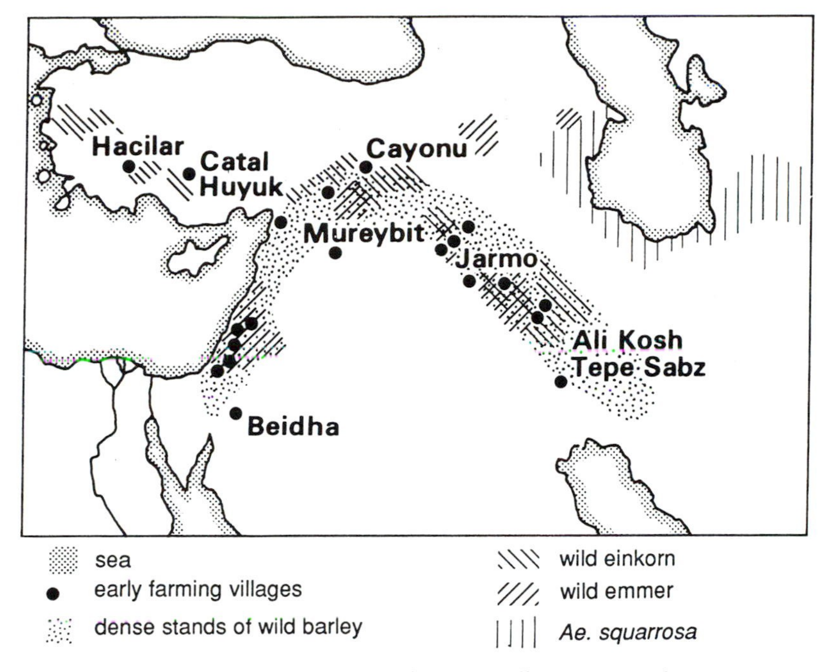


**Figure 5** The location of early farming village sites in the Near East in relation to the recent distribution of dense stands of wild barley and primary habitats of wild relatives of wheat[58].

cereals than those further away, e.g. downslope near more level, perhaps seasonally flooded, terraces.

A certain distance from the wild stands could also have been important for the domestication process itself, depending on how and when harvesting was done. Jack Harlan[82] has described how, either by hand or with a flint-bladed sickle, he harvested about 2 kilograms of grain per hour from a stand of wild einkorn wheat in south-east Turkey. From this he concluded that 'A family group ... working slowly upslope as the season progressed, could easily harvest wild cereals over a three-week span or more and, without even working very hard, could gather more grain than the family could possibly consume in a year.'

However, as Romana Unger-Hamilton[210] has pointed out, the period over which ripe grain can be harvested from wild cereal stands is often much shorter than three weeks. She knapped almost 300 flint blades which she fixed in various hafts for trials on stands of both wild progenitors and cultivated crops of several cereals and legumes. The patterns of wear on these were then compared with almost 800 neolithic and even older Natufian flint blades. She concluded that most of the older Natufian blades had been used to harvest cereals, usually at the base of the stem. Most of the Neolithic blades were heavily striated by soil particles associated with cultivation, as expected, but so too were one-fifth of the older Natufian blades. Many of these had sheen characteristic of plants being harvested green, a practice which became much less common in the Neolithic.

Given the very short period during which the harvesting of ripe grain from wild stands is possible, the cutting of still-green stems for drying and subsequent beating on ground where the seeds could be swept up would be an efficient alternative. However, it would not lead to any increase in the proportion of plants with a tough rachis, i.e. still holding their grain in the ear at maturity, the crucial characteristic of domesticated cereals.

Another of Unger-Hamilton's deductions is significant in this context, namely her conclusion that stems were mostly cut at the base. No doubt much ripe grain was harvested simply by hand or with a notched flint pulled upwards as Harlan did in Turkey and I have done above Galilee. But where fairly mature stems were cut lower down and sheaves carried back to the village for straw (bedding, matting, thatching, etc.) as well as grain, some enrichment of the seed sample with tough rachis characteristics was likely. Late harvests, for whatever reason, would have a similar effect. Nevertheless, in more than a millennium of early settlement at Abu Hureyra, some distance from stands of wild barley and einkorn, no evidence of such enrichment in plants which retain their seeds was apparent[94], despite Unger-Hamilton's deduction that pre-Neolithic cultivation was common. Indeed, many aspects of the domestication process have yet to be elucidated.

The Near East offers a particularly attractive and coherent model of domestication, plant, animal and human. There is a clearly defined centre, or rather a crescent, with many well-preserved early Neolithic village sites scattered across Anatolia, Palestine and the Mesopotamian-Zagros mountain region. At some of these we can observe the transition from hunting and gathering to the agricultural way of life. There is a wealth of well-dated botanical remains clearly showing the transitions from wild progenitor to domestic crop. Although barley and einkorn and emmer wheat were the earliest domesticates, they were soon followed by several pulses (peas, lentils, chick pea, vetch, the broad bean), oilseed, fibre (linen) and fruit crops undergoing domestication at various sites within the Fertile Crescent, where, with the possible exception of the broad bean, their wild progenitors can still be found. Animal domestication began, with sheep and goats, at about the same time as that of the cereals.

Small wonder, then, that Childe arrived at his concept of a Neolithic Revolution, and that the history of the Near East has so powerfully shaped our views on the origins of agriculture and civilization, perhaps too much so.

## 3.4 Wheat, a complex crop

The Yangzi Valley had its rice and central America its maize, but four of the world's most important cereals – wheat, barley, rye and oats – all came from the Near East. In their subsequent history they have complemented one another in the niches they have filled, both environmentally and in terms of their use in baking, brewing and animal feeding. Oats and rye were not domesticated until well after barley and wheat, indeed they were long considered as weeds.

The large grains of wild wheat and barley would have made them attractive targets for gathering and domestication, along with their more compact inflorescences than oats. Gideon Ladizinsky set his students to harvest grain from wild stands of the cereals in the Jordan Valley and found that they could gather wheat more than twice as fast as barley and four times as fast as oats. Nevertheless, barley may have been domesticated earlier than wheat, its earliest remains having been dated back to 10.2 K years ago at Netiv Hagdud in Israel, compared with 9.5 K years ago for the oldest domesticated wheat. Barley appears to have been more important than wheat through the early millennia of agriculture, but was then overtaken by it.

There are many species of wheat besides the bread wheat which is now predominant. The history of the wheats is probably best explained in terms of their evolution, which requires an explanation of what polyploidy is. Diploid species like einkorn wheat and humans have a double set of chromosomes. Tetraploids like emmer and macaroni wheats have a four-fold set, and hexaploids like bread wheat a six-fold set. Polyploidy is common among the grass family, almost three out of every four species being polyploid, although barley, rye and rice are not. The cytogenetic sleuthing to identify the three wild progenitors of wheat was begun by Hitoshi Kihara of Japan in 1919 and is still being refined.

Throughout the Fertile Crescent of the Near East and extending westwards to Greece, especially on heavy soils, there is an aggressive wild grass, *Triticum boeoticum*, whose thick stands, often mixed with wild barley and oats, can still be harvested for grain. Many carbonized seeds of this wild diploid wheat, the donor of the A genome, have been found at Near East sites in pre- and early agricultural contexts.

Then, beginning about 9.5 K years ago, along with remains of this wild wheat are found those of domesticated diploid wheat (AA), i.e. without the still-attached tell-tale segment of the inflorescence spike which signified the shattering and dispersing structure of the wild species. This einkorn wheat, so named because one race of it has only one grain per spikelet (hence *Triticum monococcum*), first appeared at several sites ranging from Ali Kosh in Iran to Cayonu in Turkey (cf. Figure 5) but the Karacadag mountains south-west of Cayonu are the most likely site of first domestication.

Often found growing with the diploid wild wheat, but over a more restricted range of environments, is a closely related grass *Aegilops speltoides*, one possible donor of the B genome of wheat. Unlike many of its relatives it is cross-pollinated and was long ago the female parent of a cross with wild wheat which may have given rise to the wild tetraploid wheat *T. dicoccoides*, or wild emmer (AABB). Cultivated emmer wheat (*T. dicoccum*) appears in the archaeological record as early as einkorn, and over as wide a range of sites. Its grains have tight glumes, whereas the free-threshing tetraploid wheats (also AABB) such as *T. durum* (for macaroni) were little known before Greco-Roman times.

The third wild progenitor, contributor of the D genome of wheat, is *Aegilops squarrosa*, whose inflorescence spike looks quite unlike that of wheat. Like wild wheat, it is a weedy grass with wide ecological tolerance, particularly for cooler climates. It was probably not until emmer wheat cultivation had spread to the Caspian sea region that *Ae. squarrosa* pollen combined with maternal emmer to create the hexaploid (AABBDD) wheats with tight glumes, such as spelt (*T. spelta*), and those with free-threshing grains such as bread wheat (*T. vulgare*). Bread wheat first appeared in the archaeological record at several places around the Fertile Crescent 7.8 K years ago. Its cultivation spread rapidly, reaching many sites in Greece within a few hundred years for example, and going on to encircle the world and occupy a wide range of environments. Polyploidy undoubtedly contributed to the broad adaptation of wheat, for example by introducing greater tolerance of cold with the D genome, which also conferred the unique bread-making quality of *T. vulgare*.

Although the evolution of wheat as a crop occurred long ago, the manipulation of its genome still creates opportunities for further improvement, as may a retracing of some of the key steps in its evolution, e.g. by the synthesis of new hexaploid combinations. The future of wheat may be quite as complex as its past.

## 3.5 The agricultures of China

Although rapid population growth did not always follow the adoption of agriculture, it did not occur without it, as Asian demographic history makes plain. The population of Asia grew 80-fold between 10 K years ago and 400 BC, mainly in those areas adopting agriculture, such as China and India. Virtually no increase occurred in those areas such as Siberia, Korea and Japan where hunting and gathering remained the lifestyle[124].

Northern India derived its agriculture initially from the Near East (p. 51) and for many years China was thought to have done likewise. Until the 1980s the earliest evidence of plant domestication was found in north-west China and included a species of millet that was also domesticated in Europe. Moreover, the five traditionally sacred grains of China included wheat and barley as well as the millets, rice and soybean. However, soybeans were not domesticated until about 3.1 K years ago, while wheat and barley entered China only about 4 K years ago, so the five sacred grains have more mythological than archaeological status.

Nevertheless, China's enviable collection of early texts and records such as oracle bones relating to agriculture offers many insights into early rural life and practices. The *Book of Odes*, dating from the 11th to the 6th century BC, mentions 150 plants compared with only 83 in the Bible and 63 in Herodotus. Irrigation is first mentioned in 563 BC in a context of opposition to it which P.T. Ho[98] interprets as indicating it was still novel. The first irrigation network, only 8 km long, operated between 424 and 296 BC. As Ho puts it, of all the ancient civilizations 'the Chinese were the last to know of irrigation'. The earliest surviving, explicitly agricultural text, the incomplete *Fan Sheng-Chi Shu* from the first century BC, includes many practical details such as exactly when to sow millet and how much seed to sow, as well as how to use silkworm excrement as an insecticide.

But whereas the historical record of Chinese agriculture is rich, access to the prehistorical record was, for a period, constrained by the Cultural Revolution. However, more than 7000 Neolithic sites have now been identified in China, ranging from north to south and east to west, but mostly focused on the Yellow River valley and the middle and lower reaches of the Yangzi.

Separated by the Qinling Mountains, the agricultures that arose in these two areas differed profoundly. In the north the early farmers settled on the grassy terraces and loess highlands which could readily be dug with their stone spades. To the south, by contrast, in lower, flatter, more humid regions, the early settlers built their villages on mounds and cleared the forest by slash and burn. Cultivation was by spades of bone and wood. Rice was their staple food whereas in the north it was millet grown after fallow periods for the conservation of moisture[118].

The earliest Neolithic cultures in the north, such as the Peiligang and Cishan, left the remains of houses, cemeteries, pottery, grain storage pits, stone spades, sickles and grain processing tools. They were followed by the Yangshao culture, between 7 K and 5 K years ago[39]. This was based on the cultivation of two millets, particularly foxtail millet (*Setaria italica*) whose probable wild progenitor is widely distributed in China, and both glutinous and non-glutinous forms of broomcorn millet (*Panicum miliaceum*). Grains of domesticated millet have been dated back to 7.8 K years ago. Indirect proof of the importance of the millets in Yangshao diets comes from the $^{13}C$ content of human skeletons, which suggests that almost 60% of the people's food came from plants with the $C_4$ pathway of photosynthesis (cf. p. 171). However, other crops, such as Chinese cabbage, jujube and mulberry, and the keeping of pigs, dogs and poultry, supplemented the diet. Hemp was also grown, and fabric woven from it, as revealed by its imprints on pottery.

Even earlier evidence of agriculture has recently been found south of the Qinling Mountains, in the middle reaches of the Yangzi Valley. Several pre-agricultural cave sites document the lives of sedentary hunting and gathering communities which relied heavily on the rich aquatic resources of the valley, and made a variety of tools and cord-marked pottery. These sites were occupied in a warm phase during which wild rice was able to spread through the Yangzi Valley. For some time the earliest evidence of domesticated rice had come from sites such as He-mu-du in the lower reaches of the Yangzi dated to 7 K years ago. In 1991 Yan[228] reported abundant remains of domesticated rice at Pengtoushan, further up the Yangzi River in Hunan, dated to 9 K–7.5 K years ago. Even older samples, with a median age of 11.5 K years, have recently been reported from the middle reaches of the Yangzi in Hubei and Hunan, suggesting that rice cultivation in China may go back substantially further than barley cultivation in the Near East.

Still further south, on the southern side of the Nan Ling mountains, there may have been a third independently arising form of Chinese agriculture which Li[118] refers to as South Asian because of its close affinities with that of the Indo-Chinese countries. Although rice is prominent in this, another cereal, Job's tears (*Coix lachryma-jobi*), was also important along with several yams and citrus species, the betel palm, lychee and other fruit trees.

China is, as Francesca Bray[22] puts it, 'the agrarian state *par excellence*'. It may well prove to have had the oldest of all agricultures, as Vavilov surmised. It encompassed the origin of several quite different agricultures, it played a crucial role in many later agricultural innovations, it has an exemplary written record of its agriculture, and it has given the world a great variety of food crops.

## 3.6 Rice, an adaptable crop

The great Swedish botanist Linnaeus named rice *Oryza sativa*. *Sativa* is Latin for sown (i.e. domesticated) and *Oryza* comes from the Greek word indicating an oriental origin. Language and agriculture have linked trails, as we shall see. Just as wheat became the staple food of the Near East and Europe, and maize of Central America, so has rice been the staple of Asia from the dawn of civilization. More than 90% of the world's rice is still produced and consumed in Asia and it is still the staple food of more people than is wheat. In the course of its history rice has been under strong selection pressures for adaptation to the wide range of Asian environments: from the equatorial tropics to the high latitudes of northern Japan; from lowlands to Himalayan terraces; from swamp and deep water estuaries to dry uplands.

The many species of *Oryza* are spread throughout the continents originally linked as Gondwanaland. Grains of several wild rice species are still occasionally gathered for food in Australia, West Africa, South America and parts of Asia. Indeed rice was domesticated not only in Asia but also in West Africa where a different crop species (*O. glaberrima*) was quite widely cultivated. Our focus here, however, is on the predominant *O. sativa*. Its probable wild progenitor (*O. rufipogon*) – not the cherished 'wild rice' of North America, which belongs to a different genus – still occurs in a broad band extending from the southern foothills of the Himalayas through northern Burma, Thailand, Laos and Vietnam to southern China. As late as 874 AD wild rice was still relied on to such an extent in parts of China that its harvesting was recorded.

There is still debate as to whether a perennial or an annual form was first domesticated, whether there was only one or several independent domestications, and where domestication occurred. For many years the oldest remains of domesticated rice appeared to be in India and Vavilov concluded that 'India is undoubtedly the birthplace of rice.' Early rice remains were also found in Spirit Cave in northern Thailand, but these are now considered to be of locally gathered wild rice. Currently, the oldest known domesticated rice comes from South China, but rice would have been a likely candidate for cultivation across such a wide region that it may well have been domesticated more than once. On the other hand, it is also argued that the swamps and marshes where wild rice thrives are such rich sources of molluscs and other food for hunters and gatherers that there would be little incentive for domestication. The grain yields of wild rice stands in both Asia and Africa can exceed 0.6 tonnes per hectare.

Rice archaeology is bedevilled, more than that of wheat, by the problem of identifying whether the charred remains are from wild or domesticated plants. The crucial early change at domestication would have been towards seed retention, which is less readily diagnosed in rice residues or potsherds than with wheat or barley. The abscission scars may prove useful, but up to now diagnoses have had to be based on grain size and glume surfaces, neither of which is really satisfactory. Of course if one finds a large cache of ancient rice, especially in storage pits, one can be fairly sure it is domesticated. The many samples with a median age of 11.5 K years recently found along the middle stretches of the Yangzi River suggest that rice cultivation began there well before the domestication of cereals in the Near East.

Further down the Yangzi many large settlements appear to have been based on rice cultivation six to seven thousand years ago. Wild rice almost certainly grew in the area then but huge amounts of rice remains, considered domesticated without doubt and of both *sinica* and *indica* races, have been recovered from settlements such as He-mu-du and Luo-jia-jiao, which were continuously occupied for more than a millennium.

The Yangzi Valley is thus a strong candidate as one site of origin of domesticated rice, without excluding others. However, both archaeological and linguistic evidence are consonant with the suggestion by Peter Bellwood[11] that the expansion of the Austro-Tai language family began among the rice-cultivating communities in coastal south China. Like the demic diffusion of agricultural peoples from the Near East (p. 48), the greater increase in the population of rice farmers *vis-à-vis* hunter–gatherers was accommodated by their progressive migration, partly by sea. Expansion through south-east Asia led to the Tai-Kadai family of languages and to rice farmers reaching Thailand and Vietnam at least 6 K years ago, and eventually as far as Madagascar. Expansion eastwards reached Taiwan 6 K years ago, and the Philippines a millennium later, followed by Borneo and Java. But whereas the Austronesian language family continued its eastwards movement across the Pacific Ocean, rice did not, being replaced by better adapted fruit and tuber crops.

Although migration eased population pressure among the rice-eaters of China from early times, improvements in the husbandry of rice crops also contributed. The introduction of the early-maturing Champa rices from Vietnam about 2.3 K years ago opened the way to double cropping in southern China[22]. This practice was then improved by the practice of transplanting rather than sowing the crops, which gave better weed control and more efficient use of land and water. The yields obtained from early rice crops are hard to estimate, more for inconsistency of units than for lack of data, but they seem unlikely to have exceeded one tonne per hectare before 1000 AD[76].

For China and for much of the world's population rice is still the staple food. Just as the spread of the Austro-Tai speaking peoples through Asia and the Pacific was dependent on the cultivation of rice so will the further improvement of this very adaptable crop be crucial to further growth in world population.

## 3.7 The Americas

About a quarter of the two and a half thousand plants that have been domesticated came from the Americas. These include important crops such as maize, potato, sweet potato, manioc (cassava) and many other roots and tubers, four kinds of bean, three squashes, many peppers, peanut, lupin, tomato, cotton, tobacco, avocado, pineapple, papaya and a great variety of other fruits, not to mention the so-called *Lost Crops of the Incas* and many productive pasture plants.

Until a few years ago, the archaeological record seemed to indicate that agriculture in the Americas began not long after that in the Near East, independently in Mexico and in the Peruvian Andes, subsequently spreading to lower altitudes in South America and, much later, into North America. However, Donald Lathrap[113] has proposed that agriculture could have begun much earlier among fisher folk living near the Amazon. New dating techniques and recent findings have made what seemed clear yesterday less clear today, to Lathrap's advantage.

One problem with the earlier picture was that if humans entered the Americas from north-east Asia during the Wisconsin glaciation and made their way south down the mid-western corridor between the ice sheets 11–12 K years ago, a view favoured by many, there seemed to be too little time for so many languages to evolve and for local populations to build up to the point where a shift to agriculture was necessary. If they walked in during an earlier lowering of the sea level by glaciation, that aspect would be less problematic for Mark Cohen's thesis that reliance on agriculture was a reluctant response to increasing population pressure.

Excavations in the dry caves of the Mexican highlands begun by Richard MacNeish[126] in the early 1960s at Tamaulipas and Tehuacan, and later at Guila Naquitz near Oaxaca, suggested a widespread early shift to agriculture. Bottle gourds seem to have been domesticated more than 9 K years ago, not surprisingly given that the earliest farmers made no pottery containers. The earliest remains of squash (*Cucurbita pepo*) and beans (*Phaseolus*), 9 K to more than 10 K years old, were probably of wild plants, but possibly domesticated remains of the grass *Setaria* and of maize, squash, beans and avocado begin to appear in contexts at least 7 K years old. The archaeological criteria of domestication in these crops are less definitive than for those of the Near East, and with maize they hinge on one's view of its origins.

Moreover, cave residues can be disturbed by rodents and human occupants over the years, so that the age of plant residues may not be the same as that of their context. This problem has bedevilled archaeologists in other regions (cf. p. 65), but is particularly troublesome in the Americas. When Austin Long[122] applied a newer dating method to some of the oldest corn cob fragments selected by MacNeish as having the most reliable provenances in two Tehuacan caves, they turned out to be about 1.5 K years younger than expected from their contexts.

Although many other crops were domesticated in Mexico, its early agriculture was dominated by maize, beans and squash. Kent Flannery has pointed out that even today in parts of Mexico, abandoned fields are invaded by teosinte, wild runner beans and wild squash. As he says, the 'triumvirate is thus not an invention of the Indians; nature provided the model'.

Further south, in the Peruvian Andes, a quite different set of agricultures emerged, with a greater dependence on root crops than in Mexico. Beans, both common (*P. vulgaris*) and Lima (*P. lunatus*) were independently domesticated there, along with other species of cucurbits, chenopods and cotton. Here too the timing is now uncertain, given that the supposedly oldest beans, from Guitarrero Cave, previously thought to be 7.7 K years old, have now been found to be much younger. However, many of the other older datings may still stand. At the higher altitudes (2.5–3.5 K metres) root crops such as the potato, oca, ulluco and many others probably predominated but left few archaeological traces. However, seed plants such as quinoa (*Chenopodium quinoa*) and, much later, maize also contributed to the diet.

Further down the eastern slopes of the Andes (1.5–2 K metres), the domesticates included common and Lima beans, lupin, peanut, *Amaranthus caudatus* and tobacco. From the Amazonian lowlands came manioc, a yam (*D. trifida*), the jackbean, peppers, squash, guava, coca, papaya, pineapple and Sea Island cotton.

It has often been suggested that the resource poverty of tropical forests precluded permanent settlement and cultural development, despite the evidence of several 19th century naturalist-explorers that there was an abundance of deep middens along the banks of the Amazon. One of these middens, at Tapirinha, was excavated in 1987 and pottery was shown to be 7–8 K years old, the oldest in the Americas and 3 K years older than the earliest pottery known in the Andes and Mexico[171]. Manioc cultivation there dated back at least 4 K years. Anna

Roosevelt and her colleagues[172] have established that the Monte Alegre palaeo-Indians were rock-painting foragers of tropical forests and rivers contemporary with North American Clovis cultures, so the Amazon forest was not the barrier to early hunter–gatherers which it has been considered.

## 3.8 Maize, the improbable domesticate

Whereas the wild progenitors of wheat and rice were attractive candidates for domestication, the likely progenitors of maize were not. Indeed, so improbable do they seem as the source of the now ubiquitous and productive corn cob that there is still a lively debate about its origins, involving many disciplines.

Detailed study of the oldest cobs from Coxcatlán Cave, and of younger ones from Bat Cave and other early rock shelters, led Paul Mangelsdorf[131] and his Harvard colleagues to propose that maize had evolved from a wild, small-seeded popcorn of the genus *Zea* which was also a pod corn, i.e. having leafy bracts enclosing each grain. This progenitor, now presumably extinct, by crossing with a related grass (*Tripsacum*), had given rise to the wild and weedy teosinte whose growth habit looks so much like that of maize but whose female inflorescence looks so different. Over the years Mangelsdorf retreated from the involvement of *Tripsacum*, but maintained that maize derived from a wild *Zea*, and not from teosinte, although he agreed that subsequent crossing between maize and teosinte may have occurred. Those earliest cave cobs were extremely small (Figure 6), with soft glumes and shallow cupules, mostly eight-rowed, and probably not wild. Mangelsdorf saw them as revealing the domestication process in action.

Of the several dissenting views, the most strongly supported is that the only wild progenitor of maize is teosinte, the vigorous grass which looks so like maize except in its female inflorescence, and which is widely distributed in Mexico. Natural populations still grow close to several of the sites where early maize cobs have been found. George Beadle[9] and others have argued that changes in only a few (3–5) genes would be sufficient to convert teosinte into a maize-like progenitor. There is considerable genetic variation among present populations of teosinte, some being annuals, others perennials, many tetraploid but

**Figure 6** A Bat Cave maize cob compared with (left) a modern dent corn and (right) a large-seeded Peruvian flour corn (P.C. Mangelsdorf[131]).

some diploid, some vigorous and widely distributed, others more localized. Molecular biological techniques have shown teosinte to be closely related to maize, and its races are now included with maize in the genus *Zea*, with the annual Mexican teosintes considered a sub-species of *Z. mays*. Perhaps, as Goodman[73] suggests, a 'name game' is being used to promote the view that maize arose from the domestication of teosinte, and to win adherents in this 'Corn War'.

However, the teosinte theory also has its problems. Unlike wild wheat and rice, it would not seem a likely or desirable candidate for domestication. Also, as Richard MacNeish[126] has emphasized, corn always occurs before teosinte in the archaeological record. DNA sequencing suggests that maize, teosinte and *Tripsacum* have all diverged from a common ancestor, with *Tripsacum* diverging first.

If the DNA base sequences of early maize cobs still have a story to tell, so do the $^{13}$carbon values of human bones. Maize operates by a different ($C_4$) pathway of photosynthesis from that in most temperate crop plants (cf. p. 171), with a lower $^{13}$carbon bias in its products. This is passed on to collagen in the bones of people eating it. The proportions of $^{13}$carbon in human skeletons from the Tehuacan Valley suggest a sharp rise in the amount of maize eaten about 6 K years ago[211]. By this criterion maize reached the Orinoco River lowlands 2.8–1.6 K years ago, and became a dominant component in the diet of mid-western Indians in North America 500–800 years ago. The more conventional archaeological evidence suggests that it may have reached South America (Colombia) about 5 K years ago, the Peruvian highlands 3.5–4 K years ago, and the Peruvian coast about 3 K years ago.

Thus, once it had become a productive crop, maize spread widely but, whatever the eventual dating of its early remains, it seems to have taken a long time for early domestication and improvement to reach the point where it was more productive of food than domesticated millet (*Setaria*) or even wild mesquite bushes. Working in Oaxaca, Anne Kirkby estimated that the threshold yield at which maize cultivation became worthwhile was about a quarter of a tonne of grain per hectare. She also measured the yields obtained there with cobs of various length and was thereby able to estimate the approximate yields of maize obtainable at various stages during the early improvement of maize. Many debatable assumptions are involved in these estimates, such as the change in number of cobs per plant as cob size increased, but the results suggest an accelerating increase in yield, with the threshold for extensive cultivation being reached 3.5–4 K years ago, about the time when the crop began to spread beyond Mexico.

## 3.9 Africa: centres or non-centre?

Harry V. Harlan, an American who devoted his life to barley, explored many countries in his search for variants of 'his' crop, but none with more enthusiasm than Ethiopia. It was natural, therefore, that he had much to discuss and debate with the great Russian student of plant domestication, N.I. Vavilov, when he stayed in the Harlan home in 1932. Harry's teen-aged son Jack was impressed by Vavilov, fascinated by the discussions and hooked by the subject. Like the two older men he has explored many aspects of plant domestication in many countries, but has particularly reshaped our understanding of crop origins in Africa.

Vavilov had been so impressed by the remarkable extent of genetic diversity in Ethiopia, in barley and emmer wheat for example, that he made it one of his eight 'centres of origin' for the crop plants of the

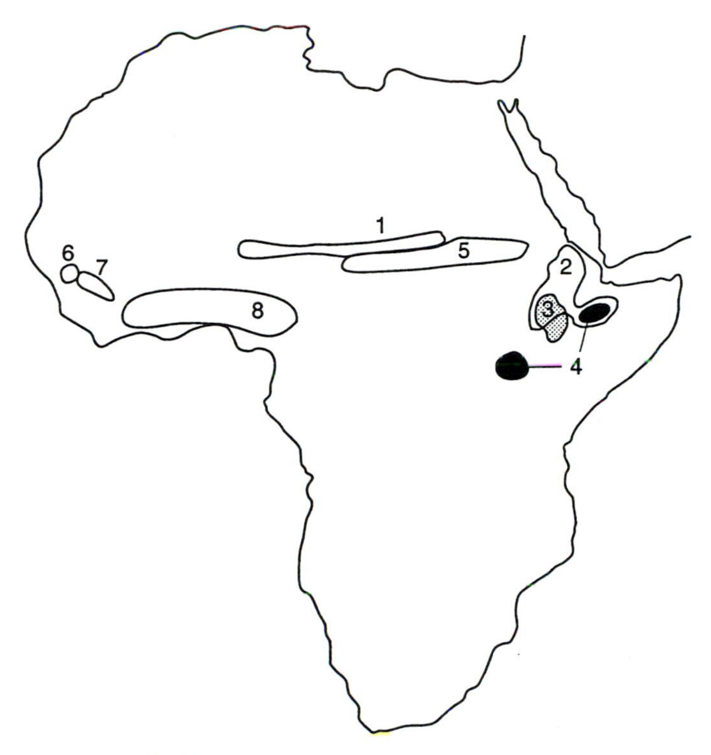


**Figure 7** Probable areas of domestication of some African crops: (1) pearl millet; (2) teff; (3) ensete; (4) finger millet; (5) sorghum; (6) fonio; (7) African rice; (8) white yam. (Adapted from J.R. Harlan[83].)

world. For him, the rest of Africa remained a dark continent. Harry Harlan was likewise impressed by Ethiopia, but he also travelled across north Africa and down into the Sahara. Writing of the latter in his autobiographical *One Man's Life with Barley*, he observed: 'There must be an area farther south where some day some collector will find many new things.' That collector turned out to be his son Jack who made six sweeps across Africa in the late 1960s, often virtually alone in remote regions, but with the advantage of motor transport. Once in the Sudan he covered by car in one day a journey that took his father 59 days by muleback.

Given the paucity of archaeological evidence of plant domestication in Africa, the probable areas of domestication of African crops must still be deduced from the present-day distributions of their most likely wild progenitors (Figure 7). This highlights the importance of a broad band

5–15°N of the equator and south of the Sahara, as Harry Harlan had intimated. While not diminishing the significance of Ethiopia it emphasizes the comparable significance of the central and western regions of the band. Jack Harlan[83] concludes: 'After many years of fieldwork in Africa and Asia I am prepared to question even the fundamental concept of 'centers' as a universal phenomenon ... The most evident contrast is between the situation in the Near East and that in Africa.'

To some extent, the apparent diffuseness of African crop origins may reflect the paucity of early dates and sites. Sorghum, almost certainly an early African domesticate, was until recently known from earlier (4 K years ago) contexts in India than in Africa. The one clear archaeological transition from wild progenitor to domesticated crop in Africa, that for pearl millet (*Pennisetum americanum*), was so rapid that it probably represents the introduction of previously domesticated pearl millet into the area about 3 K years ago, because it had already been introduced into India at least 3.2 K years ago.

Just as African crops spread to suitable environments in India, so did Near Eastern agriculture spread to North Africa and Ethiopia. Domesticated sheep, cattle and goats, together with barley, had reached the Mahgreb 7.5 K years ago, although they were apparently not adopted along the Nile until almost 1.5 K years later, presumably because there was no need to turn to agriculture any earlier.

As the Cushitic peoples spread southwards from the Near East to Ethiopia they took with them the agricultural way of life and the domesticated crops of the Fertile Crescent, especially wheat and barley. But it is still an open question whether they brought dependence on agriculture to Ethiopia or whether they encountered there farmers who had already domesticated such indigenous crops as their local cereal teff (still grown more widely there than barley), noog, ensete (the Abyssinian banana) and *arabica* coffee. The latter seems more likely. Another likely domesticate is finger millet (*Eleusine coracana*), domesticated remains of which have been dated back to 4 K years ago in Ethiopia. It too reached India at about the same time as sorghum and pearl millet.

Sorghum, the fifth most important cereal in the world, has left an unsatisfactory archaeological record. Its wild progenitor was probably *S. verticilliflorum* which grows in abundant, productive stands in eastern Africa. Harlan and de Wet[85] consider that it was probably domesticated in the Sudan–Chad area, giving rise to the most primitive *bicolor* race of sorghum through selection for seed retention, greater seed size and more compact inflorescences, possibly more than 7 K years ago. Continued selection for local adaptations led to: (1) the *guinea* race suited to the somewhat higher rainfall conditions of Sahelian West Africa and associated with the Niger–Congo language family; (2) the *caudatum* race with its distinctive grains and association with the Chari–Nile language group; (3) the *kafir* race associated with the Bantu peoples who took it with them in their surge into eastern and southern Africa in the Christian era; (4) the *durra* race, with its particularly compact heads, which appears to have been developed in India before being brought back to Africa in Islamic times.

Besides Ethiopia and central Africa, west Africa also gave rise to a characteristic group of domesticated plants. As with teff in Ethiopia, some of the cereals, such as fonio, have remained endemic, i.e. largely confined to their areas of origin (Figure 7). African rice (*Oryza glaberrima*) has been an important crop in seasonally flooded parts of west Africa, although the oldest known remains of it date back only to AD 200. Probably the most ancient domesticated plant in the more humid areas of west Africa is the yam (*Dioscorea rotundata*). Although archaeological remains are scarce and uncertain, the abundance of digging sticks for yams suggest that their culture, and possibly selection, may have predated the Neolithic Revolution of the Near East.

Thus, several distinctive agricultures may have arisen independently in a belt extending right across Africa. It is ironic that Jack Harlan, who has been referred to as 'America's Vavilov', has so broadened Vavilov's one African 'centre of origin' that it has become, in his terminology, a non-centre.

CHAPTER 4

# The first half-billion (2000 BC–1500 AD)

## 4.1 Introduction: extension, innovation and instability

By 4 K years ago, the adoption of the agricultural way of life was widespread and the population of the world was approaching 50 M. On a doubly logarithmic scale (Figure 2c, p. 6) the rate of population increase then slowed a bit. By the time of Christ, the world population is estimated to have been about 250 M, 27% in China and about 18% in each of India, south-west Asia and Europe. According to Biraben's estimates[14] the world population then remained close to 250 M for almost a thousand years (cf. Figure 2b), with drastic falls in China and India before 400 AD and in south-west Asia and Europe before 600 AD. McEvedy and Jones[124], by contrast, indicate a slow and irregular increase from 170 M in AD 1, to 265 M by AD 1000. However, both sources agree that the world population was again stationary between AD 1200 and 1400, with substantial falls in China by AD 1300 and in India and Europe by 1400. All regions were recovering by AD 1500 when the world population approached half a billion.

The drastic falls in the population of several regions which occurred periodically during the three and a half millennia suggest that agriculture was often unable to provide an assured and sustainable supply of food in the face of droughts, epidemics, wars and suppressions. Indeed, the remarkable early records of yields on the Sumerian clay discs make it clear that the productivity of the downstream irrigation areas on the Tigris and Euphrates rivers declined disastrously due to silting and

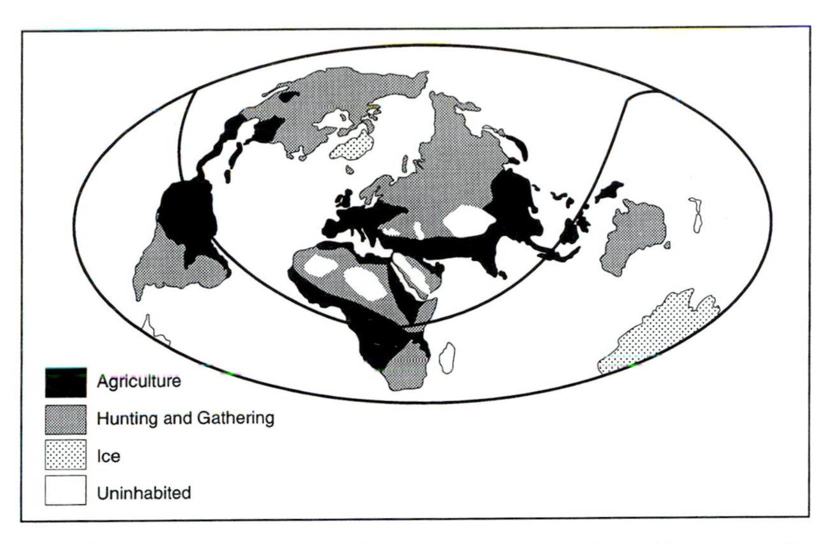


**Figure 8** Human tenure of the earth in AD 1. (Adapted from one of a series of maps by I.G. Simmons[192].)

salting. Likewise, many of the hillsides of the Mediterranean lost much of their soil from erosion during this period.

Although some high early yields of cereals were recorded both by the Sumerians and from the Nile Valley, most of the increase in food production for the ten-fold increase in population probably came from extension of the area of arable land, which was assisted by the introduction of the scratch plough or ard more than 4 K years ago. By AD 1 there was an almost continuous mid-latitude band of agriculture stretching from China and Papua-New Guinea in the east through India to western Europe and sub-Saharan west Africa and thence to central and south America (Figure 8). By 1500 AD agriculture had spread to higher latitudes both north and south. Much of this later expansion came from the clearing of forests and, to a lesser extent, from the ploughing of grassland, especially after the introduction by the Romans and Chinese of the heavy plough. In northern Europe, most of the potentially arable forest areas had been cleared by a thousand years ago.

Agriculture soon spread to the hills, aided by the introduction of terracing. Likewise, the introduction of the heavy plough, and subsequently of the horse collar, allowed the fertile bottom lands of northern valleys to be cropped, extending the arable area still further.

Swamps and wetlands were drained or converted to water meadows or chinampas, the latter often supporting large city states as in the Valley of Mexico and the Peten of the Maya.

There is much evidence, but little agreement, that the spread of agriculture from the Fertile Crescent was not by an all-conquering concept of the attractions of the agricultural way of life but by 'demic diffusion', i.e. by a wave of farming people driven outwards by population pressure. The agricultural wave advanced north-westwards across Europe at about one kilometre per year, and eastwards to the Indus Valley at about the same speed. Although much linguistic and genetic evidence supports the notion of demic diffusion[35,166], some remain unconvinced. It is also worth noting that demic diffusion from the Fertile Crescent would have been too slow for the agricultures of China and Central America, and even of south-east Asia, to have derived from that region.

As agriculture spread it also evolved. Oats and rye, minor components in the Fertile Crescent, became important crops at higher latitudes. In the Indus Valley, both at Mehargarh and later at Harappa, crops domesticated in the Near East were supplemented by later local domesticates, such as Indian dwarf wheat and cotton. And local innovations, such as the weaving of cotton, soon moved in the reverse direction. Moreover, crop plants originating in other regions often underwent secondary diversification in their new homes, as the sorghums and millets from Africa did in southern India.

With a widening portfolio of crops introduced into new environments came new needs and opportunities. The Roman two-course rotation could be changed to a three-course rotation at higher latitudes. This brought several advantages, such as greater and more stable food and then feed production, but also involved considerable social reorganization. The introduction of the heavy wheeled mouldboard plough by the Romans combined with the later introduction of the horse collar by the Chinese – an idea that quickly spread to Europe – allowed the faster, stronger horses to be used instead of oxen for cultivation. This in turn transformed not only agriculture but also the pattern of rural life, as argued so eloquently by Marc Bloch[16]. More land could be cultivated, field shapes changed, and the homes of the peasants could be further from their fields, in larger villages with more varied opportunities than in small subsistence hamlets.

Even at this early period, therefore, we begin to see the often fortuitous, but powerful, synergistic effects of separate innovations such as the mouldboard plough, the padded collar which made possible the use of horses for draft, the shift to three-course rotations and the adoption of oats as a crop. At the beginning of this period we see the slow spread of the agricultural way of life. Rather later there is the much more rapid spread of the plough, which allowed human energy to be supplemented by that of oxen. And by the end of the period we witness the rapid diffusion, both eastwards and westwards, of innovations like the curved mouldboard and the horse collar.

We also witness the profound effects of climatic change on the viability of agriculture. Just as such change set the stage for the Neolithic Revolution in the Near East, so also did it contribute to the decline of the agricultural base of several civilizations. The change to a warmer, drier climate over 4 K years ago probably reduced the flow and increased the salinity of the Tigris and Euphrates rivers, spelling doom for Mesopotamia[104]. Further east, the decline of the Harappan civilization coincided with the decline in the intensity of the south-west monsoon. The changing fortunes of the Nile and of the Aegean hills also reflect changes in climate, while the collapse of the Mayan civilization coincided with by far the driest episode in the climatic history of that region. Thus, while mismanagement of agriculture no doubt contributed to these declines, so too did climatic change.

## 4.2 The diffusion of agriculture into Europe

New techniques and concepts move around the world so rapidly these days that it is hard to grasp how slowly the agricultural way of life spread from the Middle East through Europe. It was not just that our communication systems are so much better now, because other early inventions, such as the plough or how to work bronze and iron, spread much faster than agriculture did.

The spread of agriculture from Anatolia through Europe was primarily along the Mediterranean coast to Spain or up the Danube basin into Central Europe[18]. In 1971 Ammerman and Cavalli-Sforza[4] published their interpretation, in map form, of the archaeological record of the spread of barley and wheat cultivation across Europe. Despite the irregularities of geographic barriers to migration, the earliest sites with remains of wheat or barley could be accommodated within a series of

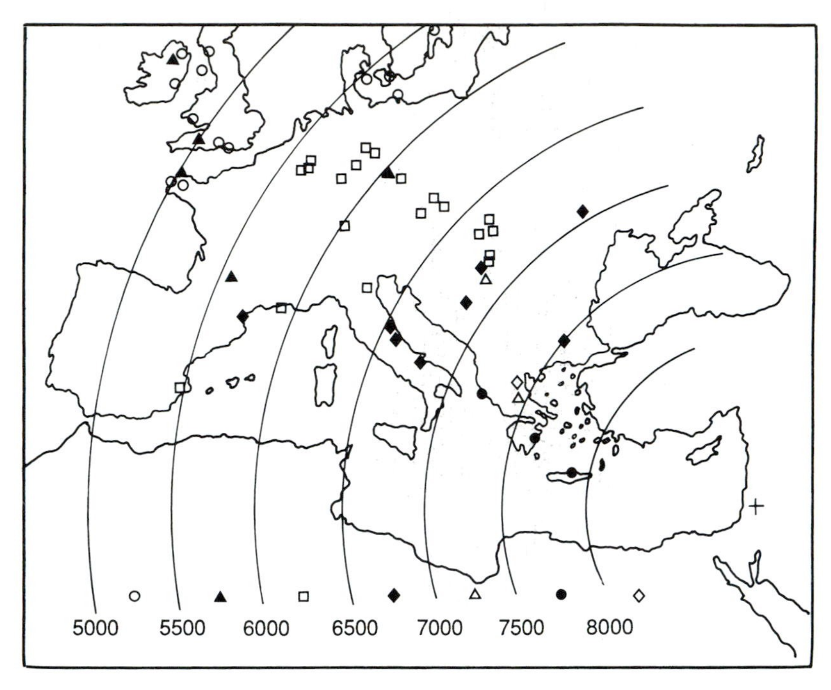


**Figure 9** The diffusion of wheat and barley into Europe. The symbols indicate the oldest remains (within a range of 500 years) found at each site. The horizontal scale is in years before the present. (Adapted from A.J. Ammermann & L.L. Cavalli-Sforza[4].)

concentric rings spreading slowly out from the Near East to reach Britain about 5 K years ago (Figure 9). But was it by the diffusion of the techniques of agriculture or by the diffusion of the farmers themselves?

Such a pattern implies a slow, uniform centrifugal extension of the agricultural way of life from the Fertile Crescent, at an average speed of less than one kilometre per year over a period of three millennia. Cavalli-Sforza and Feldman subsequently suggested that this was far too slow for *cultural* diffusion, i.e. for the adoption of new techniques or ideas by pre-existing populations. Instead, they suggested it was an example of *demic* diffusion, i.e. of a centrifugal wave of people committed to the agricultural way of life, spreading out as their numbers and need for arable land increased.

Despite much initial scepticism of their suggestion, several kinds of evidence support the likelihood of demic diffusion[35]:

(1) The major human genetic pattern in Europe is a gradient of gene frequencies radiating from the Middle East.
(2) Demographic models of population growth and migration predict rates for the diffusion of agriculture which are compatible with the observed rates.
(3) Most of the hunting and gathering peoples who have been studied are distinctly reluctant to adopt an agricultural way of life. Their training and lifestyle are very different. It seems unlikely, therefore, that the pre-existing groups of hunters and gatherers in Europe would have readily switched to agriculture. The two ways of life may have continued side by side, on separate terrain, with slow absorption of the mesolithic hunters and gatherers by the neolithic farmers.
(4) The linguistic evidence is not decisive. Colin Renfrew[166] has argued that the demic diffusion of agriculture from the Middle East could support an Anatolian origin for Indo-European languages but does not exclude other possible origins.

The demic diffusion of agriculture eastwards from the Fertile Crescent to India and southwards to North Africa also seems likely. Demic diffusion, at a slightly faster rate (1.5 km per year), may also have taken the Bantu peoples from west to southern Africa over a period of 3 K years[35].

## 4.3 Passage to India

In 1926 when Nicolai Vavilov focused attention on just eight 'centres of origin' for the crop plants of the world, he made the Indian subcontinent one of the eight and concluded that 'India is undoubtedly the birthplace of rice.' As we have already seen, however, the oldest known remains of domesticated rice come from the Yangzi Valley in China.

The earliest signs of agriculture in the Indian subcontinent come from the Indus Valley in the north-west. There at Mehargarh, on an alluvial fan at the foot of the Bolan Pass, one of the two main routes between the Indus Valley and the Iranian plateau (Figure 10), evidence was found by a French archaeological team of a transition from dependence on hunting and gathering to agriculture possibly 3 K years before the Harappan civilization further east.

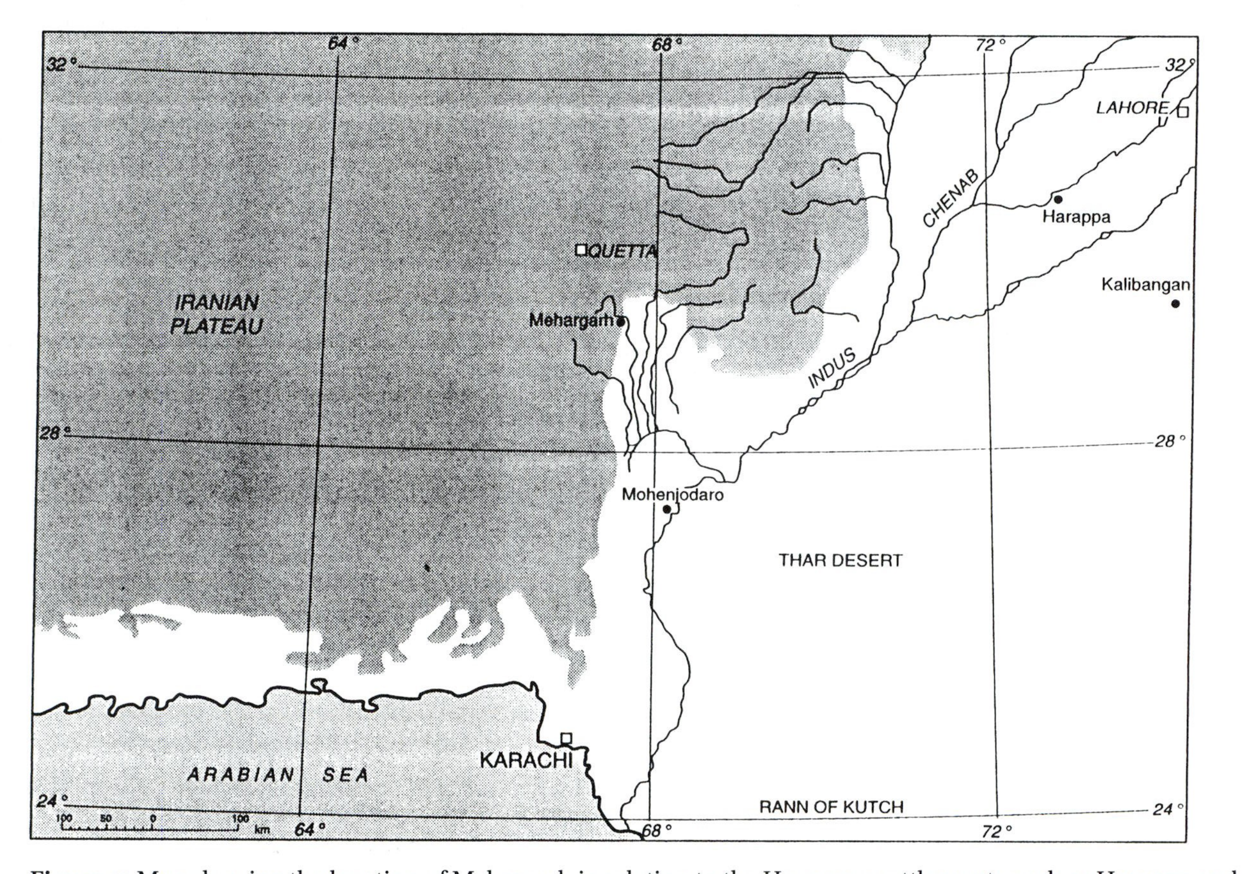


**Figure 10** Map showing the location of Mehargarh in relation to the Harappan settlements such as Harappa and Mohenjodaro. (Adapted from R.H. Meadow[132].)

Jean-François Jarrige and Richard Meadow[107] suggest that the Mehargarh settlement began more than 8 K years ago. Within the lower layers of the deposits there is a clear shift from the bones of wild animals such as gazelles, deer and wild cattle to those of domesticated cattle, goats and sheep. Since the wild relatives of all these domesticates roamed Baluchistan over 7 K years ago some, particularly the sheep, may have been domesticated locally during the transition from hunting to herding. However, the plant remains at Mehargarh are strongly suggestive of an agriculture derived from the Fertile Crescent of the Near East. Impressions of both two- and six-row barley, einkorn, emmer and durum wheat, jujube, grapes and date palm have been identified among the early deposits. Pottery first appeared about 5.7 K years ago.

Over the years, however, the domesticates brought from the Fertile Crescent were supplemented by the domestication of local plants. Seeds and fibres of cotton, probably of the perennial Indian tree cotton (*Gossypium arboreum*), appear at Mehargarh more than 6 K years ago. Grains of Indian dwarf wheat (*Triticum sphaerococcum*) appear more than 5 K years ago, associated with the spread of agricultural settlements in the Indus Valley.

The archaeological scene then shifts eastwards to Harappa and down the Indus Valley to Mohenjodaro (Figure 10) and other sites excavated in the 1930s by Sir Mortimer Wheeler. These represent the Harappan civilization which extended from about 4.3 K to 3.6 K years ago and was characterized by elaborate buildings and town planning. Here too we see clear signs of Fertile Crescent agriculture, with 4 K year-old remains of wheat, barley, peas, lentils and chickpeas, plus evidence of ploughing more than 3 K years ago.

The oldest known remnant of cotton cloth, probably woven more than 4 K years ago, comes from Mohenjodaro, and it is possible that this innovation moved in the direction opposite to the food staples, i.e. from the Indus to the Fertile Crescent. Such characteristic Indian domesticates as green and black gram (*Vigna radiata* and *V. mungo*) and various mustards are also found, as well as rice.

In southern India, by contrast, the characteristic element seems to have been an early influx of crops domesticated in north-east Africa. Finger millet (*Eleusine coracana*), so characteristic of Indian agriculture that de Candolle and others thought it was domesticated there, almost certainly came from Africa where its Arabic/African name of telbum is synonymous with the Nubian word for cultivation. Sorghum, bulrush millet, *Dolichos*, cowpea and other African domesticates also reached India, several of them more than 3 K years ago, via an unknown route.

As always, fewer traces of early agriculture have been found in the wetter regions of the subcontinent. Crops such as Indian hemp (*Cannabis sativa*) came in from Central Asia along with other northerly crops such as cucumbers from the Himalayas. Rice and other crops such as the mango are quite likely to have been domesticated in north-east India. Many south-east Asian domesticates are also likely to have entered through the north-east, such as sugar cane, which written records suggest was widely cultivated on the Ganges plains at least 3 K years ago. The banana, giant alocasia, taro and other south-east Asian crops also probably entered India long ago.

Thus, although some Indian domesticates are among the world's most important crops today, while many others contribute variety and flavour to our diets, it was the passage to India of crop plants from the north, south, east and west that bestowed a remarkable diversity on Indian agriculture and food. Many of these introductions, such as those from Africa, underwent secondary diversification in India to such an extent that they have been considered Indian in origin. Yet the limits to such local variation and adaptation are also clear in that the species brought in from the winter rainfall areas of west Asia are still predominant in the west, those from the summer rainfall areas of Africa still characterize the south, and those of south-east Asia dominate the more humid easterly regions. The wealth of secondary variation in India has not changed the basic adaptive responses of the crops from elsewhere.

## 4.4 Pastoral nomadism and the horse

Nomadic pastoralists and sedentary agriculturists had, since long before the Scythians and the Mongols, disliked, distrusted and made war on one another. Only in the Near East was the domestication of animals as prominent and as early as that of plants for agriculture, and even there the relation between them at the beginning is unclear.

In the early 19th century it was widely believed that mankind evolved from hunting and gathering through a stage of pastoral nomadism to sedentary agriculture. Then it was suggested that pastoral nomadism developed as an offshoot of crop cultivation. Currently it appears that sheep and goats were domesticated within the Fertile Crescent of the Near East 10–11 K years ago, at about the same time as wheat and barley. At Beidha (cf. Figure 5, p. 27), for example, the earliest stages of barley domestication coincided with the early domestication of goats. Domesticated pigs date back to 8.8 K years ago at Cayonu in Turkey, and domesticated cattle back to 8.5 K years ago. Domesticated cattle and pigs did not appear in south-east Asia until about 5 K years ago, and mitochondrial DNA studies suggest that Indian cattle and, later, African cattle had separated from Near Eastern cattle long before, i.e. that they were independently domesticated.

Herding animals thus appear to have been domesticated by sedentary agriculturists, not by hunter–gatherers. Charles Reed[165] has suggested that at least the social bovids (cattle) were pre-adapted to domestication and that what was really needed was a change in human attitudes from hunting to herding. However, it would have been necessary to keep the herded animals well away from the primitive crops, at least until after harvest. Then, as local populations increased, it is likely that shepherds had to go further and further afield in search of pastures, becoming specialized herders whose contact with the agricultural villages became more tenuous. Most animals, even dogs, were probably domesticated as a source of meat. Evidence of milking appears only about 5 K years ago, and the processing of wool began well after initial domestication, as did the use of cattle as draft animals.

However, it was the domestication of the horse about 5 K years ago that made nomadic pastoralism a dominant way of life. The wild horse, *Equus przewalskii*, roamed from western Europe to eastern Asia, but it appears to have been first domesticated in the Kuznetz steppe of southern Russia, initially as a source of meat. It soon spread westwards into Europe and south to the Near East by 4 K years ago, where horses were used to draw chariots. The riding of horses, which became common in the Near East only about 3.4 K years ago, made the herding of stock over long distances much easier, particularly after the invention of the stirrup, first illustrated in India in the second century BC but possibly of Chinese origin[222].

Horse-riding skills also gave the pastoral nomads military advantages over sedentary agriculturists. From the rise of the Scythians in the Volga region 2.9 K years ago to that of the Mongolian herdsmen, pastoral nomadism dominated much of the drier regions of Eurasia and North Africa. In fact, early attempts at agricultural development of the steppes, e.g. in Russia from the second century BC, were rebuffed by the pastoral nomads until the 17th century AD[77].

In the more arid regions of the Near East and North Africa the horse was replaced by the dromedary, domesticated about 4 K years ago in south-west Arabia, but the antagonism between farmers and pastoral nomads was still apparent. Nevertheless, many nomads retained links with agricultural communities, and would even sow barley or millet in recently flooded wadis on their way out to spring pastures for harvest during their return. Women and children might even be left at water holes to tend such crops, tempting even the mobile nomads with the advantages of a more sedentary lifestyle.

## 4.5 The plough

Like the digging-stick and the almost universal hoe, the plough may have needed only human power initially, but it soon involved the first application of non-human power to agriculture. As such its use spread rapidly through the old world, but its origins remain obscure. In the 1970s the Danish archaeologist Axel Steensberg[200] found several triangular wooden spades with long handles and two holes at the top of the blades for pulling by ropes, in Satrup Moor, Schleswig. Their context dated to 6.2 K years ago. Some years earlier he had found similar tools but made of basalt or limestone in Syria, dating back 4.4 K years.

The ard is a light, usually wooden, plough of many shapes, well suited to the sowing of cereals in lighter soils and drier conditions, and requiring only one or two oxen for draft. Indeed the ard could still be seen in use around the Mediterranean and in Asia until recently. The oldest known ard, or scratch plough, in Europe was found in a bog in Jutland and dates to 3.5 K years ago. However, there are much earlier traces of ard ploughing dating back 5.6 K years in Sarnawo, Poland, while the earliest picture of an ard plough from Babylonia, with a beam and sole drawn by two oxen, dates to 5.2 K years ago. A combined plough and drill from early Babylonia is illustrated in Figure 11.

So much power is required for ploughing as against hoeing that use of the ard probably became widespread only in conjunction with the availability of a domesticated draft animal. Of the early farm livestock only oxen were sufficiently powerful, and the earliest evidence of their domestication comes from Greece about 8.5 K years ago. In view of the

**Figure 11** Babylonian ard with attachment for drilling grain, 3300 years ago (R.H. Anderson[5]).

long period between then and the first appearance of the ard, its invention could have occurred anywhere in the Near East–Balkans region.

Once it had made its appearance, however, it spread rapidly, much more so than did the agricultural way of life. Ard furrows dating back 4.4 K years have been found in north-west India and cave paintings of ards in Sweden date from more than 3 K years ago. Although many scholars have suggested that the ard did not reach China until 500 years ago, it may have been in use there since the fourth millennium BP, judging from stone plough shares found near Shanghai[22].

In more northerly regions the ard could not cope with the heavier, more fertile soils of the valley bottoms, where heavier ploughs were needed. The mouldboard or turn plough was not used until more than 3 K years after the ard was introduced, appearing in Europe in the first century AD and in China by about 700 BC[22]. Pliny contrasted the light ploughs used in Syria in the first century AD with the heavy wheeled plough drawn by eight oxen then being used in the Po Valley but supposedly invented in the foothills of the Italian alps. The addition of wheels made the plough more mobile, and helped the ploughman to adjust the depth of his furrow. Pliny also mentions the coulter, the heavy knife which cuts the sod vertically down to the ploughshare.

The other new component was the mouldboard which turned the sod over. In Europe this consisted of a flat wooden structure until the 17th century. In China, by contrast, curved iron mouldboards which married smoothly to the surface of the iron shares were in use by the first century BC, conforming in many respects to the principles of mouldboard design first enunciated in Europe in the 18th century by James Small of Scotland[22]. One consequence of this difference was that larger teams of draught animals, and relatively more pasture and feed, were required in Europe compared with China. On the other hand, Chinese ploughs had no wheels and no coulter, indicative of the independent origins of the mouldboard plough in China and in Europe.

The invention of the horse collar in China made possible the use of horses for heavy draught work, permitting faster, more timely and more distant cultivation. The throat-and-girth harness first illustrated by the Chaldeans 5 K years ago was universal and unchanging. While adequate for oxen, it tended to suffocate horses and, as a result, only a fraction of their full power could be used, enough for a cart but not for a heavy plough. Hence the Theodosian code of 438 AD which decreed that anyone harnessing horses to a load of more than 500 kilograms would be severely punished. Yet only one civilization, namely China, solved the problem with the introduction of the padded collar harness, so well adapted to the anatomy of the horse. Joseph Needham[146] suggests that this first appeared in China about 475 AD. The earliest illustration of it in Europe was about 920 AD, and it was universally adopted in the West by the end of the 12th century AD.

The invention of the scratch plough over 5 K years ago transformed agriculture in terms of the power available for cultivation, a major limitation on the area that could be sown. The invention of the mouldboard plough three millennia later also had a major impact on agriculture, particularly by making possible the cultivation of the more fertile and well-watered heavy valley-bottom soils, especially at the higher latitudes. Also, by eliminating the need for cross-ploughing, the heavy plough saved time and made the cultivation of long, narrow strip fields more practicable, improving drainage as well as influencing patterns of inheritance.

The French historian and Resistance hero Marc Bloch[16] has argued that the heavy plough had even more profound effects in shaping peasant societies in northern Europe. The eight oxen needed to pull it in heavy soils were beyond the resources of most peasants, who therefore had to pool their teams. Their cooperation extended beyond this, however, because the ploughing and planting of their various strips had to be coordinated, giving rise to the need for strong village councils in the manorial economy of northern Europe. In his 'grand hypothesis', as White[222] refers to it, Bloch argued that these are only intelligible in terms of the heavy plough.

The introduction of the padded horse collar also had profound effects on agriculture, by allowing the greater strength, endurance and mobility of horses to be utilized. Cultivation became more timely, the farmers could live further from their fields and therefore in larger villages with more diverse opportunities, while the sale and faster transport of surplus produce encouraged commerce and the growth of cities.

## 4.6 Sumerian grain yields

To the later Romans, such as Herodotus, the yields of wheat and barley in Mesopotamia 4 K or so years ago were the stuff of fable. But were they? By what criterion of yield? And how can we know?

Having invented non-pictorial writing over 5 K years ago, the Sumerians became compulsive record keepers and many accounts of cereal harvests, field by field, have survived on clay discs, like the one illustrated for barley in the year 2039 BC, during the third Ur dynasty (Figure 12). To what extent the recorded yields are representative of the agriculture of the time, or refer only to centrally-controlled farms, is not known. There are also uncertainties about the conversion of units for grain volume and land area to present day ones, e.g. from gur per bùr to litres per hectare, especially with units which changed over time. And there is the likelihood that the weight of wheat and barley per litre of grain was different from present day values. So much scholarship has been devoted to these problems, however, that the estimated average yields are likely to be reasonably sound.

**Figure 12** Sumerian clay disc listing fields and their yields of barley in the year 2039 BC. (Courtesy of the Trustees of the British Museum.)

They tell an interesting story about siltation and salination. Agricultural settlement between the Tigris and Euphrates rivers – i.e. in Mesopotamia, 'the land between the rivers' – began about 6 K years ago. By 4.4 K years ago there were well-developed principalities such as Girsu and Umma with smaller settlements aligned along the water-courses and using some form of irrigation for their crops. Girsu at that time seems to have produced more or less equal amounts of wheat and barley, judging by the frequency of impressions of their grains on pottery sherds. The average yields recorded about 4.4 K years ago of 2537 litres per hectare would be equivalent to about 2 tonnes per hectare of wheat, or 1.5 of barley. Such an average wheat yield per hectare was several times greater than Roman yields, and was reached in England only around 1900 AD, so it was remarkable.

However, grain yields in Roman times were usually compared in terms of the ratio of volume of grain harvested : volume sown. The yield ratio for Sumerian barley crops was commonly 30-fold and possibly up to 76-fold in the earlier years[162]. Such ratios were very high by Roman standards. But, in spite of many statements to the contrary, they can be matched, indeed surpassed, by high yielding modern crops.

One reason for the high yield ratios was the apparently low sowing rate, and therein lies a clue. It seems likely that the early Mesopotamians sowed the seed sparingly but carefully in widely-spaced furrows which may also have served to guide the irrigation waters. The well-watered and well-spaced plants that resulted would have tillered profusely and borne many ears, to establish a high yield ratio and a long-lived legend of exceptional soil fertility.

It was not to last, however. Average yields in the Girsu area had fallen to 0.8 tonnes per hectare by 4.1 K years ago, i.e. they were halved in a period of 300 years. And by 3.7 K years ago they had fallen to half a tonne per hectare, by which time many of the Sumerian cities had declined to villages or ruins, while Babylon rose and political dominance passed to cities further up the river from the accumulating silt and salt.

No doubt many elements contributed to the decline in the yield of cereals. The archaeologists Thorkild Jacobsen and Robert Adams[106] have suggested that progressive salination of the fields occurred, beginning soon after the ruler of Girsu, Entemenak, built a large canal to bring water for irrigation from the Tigris about 4.4 K years ago. They found temple records attesting to the appearance of salty patches in fields around Girsu soon after Entemenak's reign. In a few cases the records showed that individual fields which had been previously recorded as salt-free were no longer so by 4.1 K years ago.

As further evidence of salination, Jacobsen and Adams cite progressive replacement of wheat by the more salt-tolerant barley, as judged by their proportions in the impressions on pottery sherds. From being about half of the markings made 4.4 K years ago, wheat had fallen to 2% by 4.1 K years and to virtually nil by 3.7 K years ago. Along with this shift, ale became a more important element of the diet relative to bread.

Although salting of the soil, possibly through more prevalent overwatering once the Tigris canal was built, undoubtedly contributed to the fall in cereal yields, many other factors may also have done so : a change in climate, loss of soil fertility after prolonged cropping, the silting up of canals leading to less reliable irrigation, or a decline in the power of the central government.

Thus, while irrigation undoubtedly made possible the high yields of cereals in early Sumerian agriculture, its mismanagement was not necessarily the only, or even the major, cause of their subsequent fall, which paralleled that of the Sumerian cities.

## 4.7 Terracing the hills

Although agriculture may have begun in the lowlands, population and other pressures soon led it into the hills, where the spread of farming was soon associated with the development of terrace agriculture. Just as the permanence of agriculture on the flood plains depended on the nature of the rivers and their burdens of salt and silt, ranging from the perennial fertility of the Nile to the shorter lived fertility of Sumer, so has the permanence of hillside agriculture depended on climate, soil and social organization. Throughout much of the Mediterranean region the mountains have been denuded not only of forests but also of soil, whereas in South America and Asia many of the cultivated terraces are remarkable not only for their scale but also for their longevity.

In the cooler, wetter areas of northern Europe where the hillside forests were replaced mainly by perennial pastures for grazing, stable farming systems developed. But in the Mediterranean regions where heavy autumn rains followed a long dry season on soils of limited storage capacity, the clearing of the hillside forests for cultivation often led to heavy run-off, soil erosion and an accelerating cycle of water, soil and nutrient loss from the hills coupled with excessive accumulation in the valleys as marshy, malarial swamps[93]. Archaeological research in Greece is showing that over the past 7 K years the periods of intense human settlement correlate with the periods of erosion, which have in turn often led to the eventual abandonment of the settlements[178]. Much of the damage was done in the Hellenic period, beginning about 2.8 K years ago. Like the cedars of Lebanon before them, the forests were cleared for timber, the land was ploughed and the cycle of erosion began.

The sad fact is that the Greeks, so often cited by environmentalists for their reverential attitude to nature, recognized the damage they were doing. Even 2.8 K years ago, Homer referred in the Iliad to the tilled fields of men being wasted by the torrents rushing headlong from the mountains. Solon later recommended the cessation of cereal cultivation on the Attic hillsides. Five hundred years after Homer, Plato summed up the situation in one of his Dialogues: 'What now remains of the formerly rich land is like the skeleton of a sick man ... Formerly, many of the mountains were arable. The plains that were full of rich soil are now marshes. Hills that were once covered with forests and produced abundant pasture now produce only food for bees. Once the land was enriched by yearly rains, which were not lost, as they are now, by flowing from the bare land into the sea. The soil was deep, it absorbed and kept the water in the loamy soil, and the water that soaked into the hills fed springs and running streams everywhere. Now the abandoned shrines at spots where formerly there were springs attest that our description of the land is true.'[93]

So here was an articulate civilization in the throes of expansion, well aware of what it was doing to its own countryside yet seemingly unable to moderate the demands of its growth in population and power. Likewise, a little later, in the Apennines of Italy. Thus, 'the glory that was Greece and the grandeur that was Rome' were built on, and eventually undermined by, the denudation of their hills and mountains. Even distinguished biologists have been known to revel in 'the divine nudity' of the Aegean hills, but had these remained clad in their former forests and soils, their recreational value and ameliorating effects on water supplies and the environment would have been cherished even more.

The technique of forming cultivable terraces by building walls of stones or earth along the contours of hills is at least three and a half thousand years old and may have developed independently in several regions of the world given its variety of form and purpose. In areas of seasonal rainfall, terraces helped to conserve both soil and water. They could also concentrate run-off water for the irrigation of perennial crops, as in the Negev[60]. In Asian rice growing regions the network of mud walls topped with turf running ladder-like up hillsides created the requisite paddies for puddling and weed control as well as nutrient conservation. The earliest known terraces in China date back to 2.2 K years ago, but they were not common before the 9th century AD[22]. Although the 'high-rise' rice terraces of Asia are often claimed to be of great antiquity, they may not be so. The famed Ifugao terraces at

Banaue in the Philippines are probably less than a thousand years old. The equally-famed 'andenes' of South America, with their carefully constructed stone walls which go back at least 2.9 K years near Lake Titicaca, were widely used in the highlands of Peru by AD 300 and expanded greatly under the Incas, particularly for the irrigation of maize[143].

The large-scale terracing of Andean and Asian hillsides required sustained cooperation among those working and maintaining the terraces at various levels, over many generations. Neglect could quickly lead to the failure of some terrace walls leading in turn to the breaching of others lower down. The hills of Asia and South America are living testimony to the value of terrace-building in the creation of agricultural landscapes that are both attractive and of sustained productivity. Equally, the many abandoned and eroded remains of terraces throughout the Mediterranean, as well as the 'divine nudity' of the hills, bear witness to the destructive power of unwise cultivation.

## 4.8 Chinampas and the collapse of the Maya

Chinampas have been variously referred to as raised, ridged, wet or island beds, floating gardens, or drained fields. 'Floating gardens' is certainly a misnomer, originated in 1590 by Father Acosta in his *Natural and Moral History of the Indies* when he confused the towed rafts of water plants to be used as compost on the raised and fixed beds with the chinampas on the Lake of Mexico, one of the best known examples of them.

However, the building of raised beds for the cultivation of crops in flooded or swampy areas has been practised for so long in so many regions – Central and South America, Africa, south-east Asia, China, New Guinea and the Pacific[53] – that they presumably had several independent origins, and my use of the term 'chinampa' is not intended to ascribe their invention to the Americas alone. Indeed, the oldest trace of them might be at the Kuk swamp in the highlands of New Guinea about 9 K years ago, but they certainly date back 4 K years in the Guayas basin of Ecuador. In the Mexican highlands, the Peruvian Andes and the lowlands of the Mayan civilization, chinampas were in extensive use about 2 K years ago.

What was the reward for the huge effort of digging a network of channels and using the excavated soil to build a grid of beds up to 150 metres long and 5–10 metres wide, a metre or so above water level, with protective and supporting walls of sods, woven brush, stones and shrubs?

Swamps, lakes and rivers often attracted hunter–gatherers because of the wealth of their food resources, including molluscs, fish, turtles and birds. Seasonal flooding is associated with renewed soil fertility, but also with a frequent need for water management and land reclamation. Many swamp soils are not only fertile but relatively easy to cultivate with digging sticks and primitive spades. In seasonally-dry areas like the Mexican highlands they provide assured access to water, the major constraint to crop growth. Particularly in the higher regions, like the Peruvian Andes, both the winter and summer extremes of temperature are moderated by the water. These advantages combined to make chinampas highly productive agricultural systems, often with the additional advantage of efficient transport of their produce by canoe or boat.

Like the irrigated agriculture of Mesopotamia and the Nile and the wet rice culture of China, chinampas made possible the support of large urban populations and their development of specialized crafts and social hierarchies. In the seasonally-arid Valley of Mexico they supported the Aztec capital where at least 100 K people lived[43]. The cultivated area of chinampas on just two (Chalco and Xochimilco) of the five lakes in the Valley has been estimated to have been 10 K hectares, and the labour involved in their maintenance and in their continual cropping with maize, beans, peppers, tomatoes, amaranthus, cotton, vegetables and flowers with only digging sticks must have been immense. Yet the system survived the Spanish conquest in 1519, and is used to a small extent even today in Xochimilco.

Thus, chinampas can apparently combine high productivity with sustainability, although the sudden collapse of the Mayan civilization about 800 AD, with a fall in population from 3 M to 0.5 M, leaves a shadow of doubt. As with the Aztecs, the extensive reliance of the Maya on chinampas, which at one stage occupied over one-fifth of the land surface in the Peten area[192], had been coupled with a marked increase in population and power. Their earlier reliance on shifting cultivation in the pre-Classic era extending from more than 4 K years ago to AD 250, was not sufficient for urban development, yet probably remained the main source of food production in the Classic era up to 800 AD. Consequently it is not clear what the sudden collapse of the Mayan civilization at the height of its power was caused by and why rich agricultural land which had been cultivated for a millennium was abandoned.

The collapse began in the most populous central region, but overpopulation may not have been the cause. With the Irish potato famine and the 1970 epidemic of maize leaf blight across the USA in mind, James Brewbaker[23] suggested that the Mayans had become too dependent on the continuous culture of maize, particularly on one variety, Nal-Tel, with the aid of chinampas. Pest and disease outbreaks could have become serious in such conditions, and Brewbaker suggested that an epidemic of hopper-borne mosaic virus from the Caribbean could have caused the Mayan collapse.

As always with early civilizations, however, other causes may have been operative, and have been proposed. One of these was a change in climatic conditions, often suggested but lacking strong evidence until David Hodell and his colleagues[99] examined oxygen isotope changes in the water and sediments of Lake Chichancanab in the Yucatan. Their results showed that the collapse of Classic Mayan civilization coincided with the onset of the driest period by far in the climatic history of that region over the last 7 K years.

Thus, the Mayan collapse was probably not due to either overpopulation or inherent instability of the extremely intensive chinampa system of food production, but to a fairly rapid deterioration in climatic conditions in much of the region beginning about 800 AD.

## 4.9 The fabled Nile

In 1978 Fred Wendorf and his colleagues found four carbonized grains of barley in a buried hearth at Wadi Kubbaniya, just north of the Aswan High Dam in Egypt. These grains generated a lot of excitement because the context in which they occurred was carbon-dated to between 17.1 and 18.3 K years ago, yet the dimensions and other features of the grains were like those of domesticated barley. If valid, such a finding would almost double the time since the beginning of crop domestication in the Near East and make the Nile its home.

Wendorf and his colleagues had considered several possible problems with this dating, such as whether the climate would have been suitable for growing barley there at that time. But they temporarily overlooked a crucial clue, which was that although the grains were carbonized, they had not been charred sufficiently to be likely to survive for so long. Several years later they discovered some more barley seeds in nearby and similar contexts. Using a more elaborate dating technique they found that a grain from the same context as the earlier ones was indeed only 4.8 K years old[217]. Clearly the four grains had intruded into the older context.

This was not the first time such problems have bedevilled archaeology. Wendorf and others had long pointed to the abundance of grinding stones, handstones and bladelets at various upper Nile sites dating back to 15 K years ago. Because grinding stones are extremely rare in the Palaeolithic of the Near East, their abundance in Nubia 15 K years ago was widely assumed to indicate a beginning of mankind's dependence on seed harvesting for food in the Nile region. More recent archaeological work at Wadi Kubbaniya has shown, however, that so-called 'root' foods, such as the tubers of nutgrass, were more common than grains in the diets of the time. When roasted to remove toxins, these become so hard that they need grinding, hence the many grinding stones[95].

The upper Nile, therefore, was not a step on the way to the agriculture of the Fertile Crescent. Nor did that agriculture reach the Nile until about 6.5 K years ago. This brings yet another unexpected twist in the story of the Nile, namely that forms of agriculture based on barley, cattle, sheep and goats – clearly derived from the Near East – appeared at least 1.5 K years earlier in the Western deserts than they did along the Nile.

One explanation of the late adoption of agriculture along the Nile is that the Nubian hunters and gatherers already had such a reliable and productive economy that they had become culturally conservative. Another is that until the post-glacial rise in the level of the Mediterranean sea had slowed down sufficiently for the Nile silt to accumulate in the delta, settled agriculture would have been difficult, i.e. until at least 7.5 K years ago[199]. Only then could the black soil (khami), from which Egypt (Khemia to the Greeks) and chemistry took their names according to Hillel[93], begin to accumulate.

In yet another twist in the Nile story, and surely not the last, Wendorf and his colleagues[218] have found evidence, 100 km west of Abu Simbel, of the intensive use of sorghum and several millets, legumes, mustards and jujube 8 K years ago. Their striking, but still uncertain, finding was that, in some respects at least, the sorghum grains resembled those of

**Figure 13** The Scorpion King inaugurating an irrigation network, *ca.* 3100 BC (from K.W. Butzer[30]).

domesticates. A Nile agriculture, independent in origin from that of the Near East, would be implied.

Agriculture along the Nile quickly developed its own forms and fables. The earliest evidence of irrigation anywhere is the mace head of the Scorpion King inaugurating an irrigation ditch about 5.1 K years ago (Figure 13), as interpreted by Karl Butzer[30]. The Nile has remained, at least until recently, one of the best examples of permanent irrigation agriculture, in sharp contrast to the more silty and salty flood plains of the Tigris and Euphrates, and reflecting differences in the wet season at their headwaters and in the width of their floodplains[93].

In Roman times, when Egypt supplied a third of Rome's grain, its yields of wheat were legendary, 10–27-fold higher than the amount sown (i.e. up to 2 tonnes per hectare), compared with an average of 5–6-fold in Rome. Even in 1798, when Napoleon's team of savants surveyed Egypt in great detail, crop yields per unit area were still outstanding, exceeding those of western Europe[198]. Egyptian yields are still high, although no longer the highest, and agricultural production still depends almost exclusively on the Nile. The fabled river still has much to offer, not least in its temptations to archaeological imaginations and the twists and turns in their unravelling of its history.

## 4.10 The maintenance of soil fertility

Compared with the philosophical bent of the Greek writers on agriculture, the Roman authors from Cato to Columella grounded their writings more firmly in their practical farming experience. Their lives coincided with the decline of the independent peasantry during the late Republic, and their experience derived from the large, slave-operated estates, the *latifundia*.

The earliest of them, Cato the Censor (234–149 BC), was a champion of the common people and the simple life. His *De Agri Cultura* is a rather rambling account of traditional practices. Varro (116–27 BC) began his *Res Rusticae* in his 80th year. Virgil (70–19 BC) derived much of his agricultural knowledge from Varro's three books, but it was his *Georgics* which carried Roman practices to the outlying parts of the Roman empire and through the Middle Ages.

Finally there was Columella (~ 4 BC–AD 65), whose *Res Rustica* in twelve books was the most comprehensive and systematic of the great Roman treatises on agriculture. Columella begins his work by denying the apparently widespread view that soils, like people, necessarily become worn out with age and over-production, in the extreme becoming *agri deserti*, i.e. deserts. Rather, he says, their loss of fertility is due to the formerly careful husbandry of the Roman farmers being handed over to the *latifundia* slaves 'as if to a hangman for punishment; whereas for oratory, mathematics, music, shipbuilding or war one goes to experts for training, the most important art of all, agriculture, is as destitute of learners as of teachers.'

Columella's treatise therefore offers detailed practical advice on all aspects of agriculture: how to yoke and handle your oxen in ploughing, always across the slope by the way; when, where and how to sow the many crops and their varieties; the uses of various implements and types of drainage; the number of labourers required, etc. A dominant theme throughout is the maintenance of soil fertility, 'quickening the earth', by the use of animal manures, green manures and crop rotations.

The traditional practice on lighter soils throughout the Mediterranean region was to allow the land to lie fallow for a year between autumn-sown wheat crops, partly to conserve moisture and partly to restore soil fertility. Cato was aware of the beneficial effects of crops of lupins, broad beans and vetch on soil fertility. Varro comments that besides forage crops such as clovers, vetches and lucerne, the legumes (so-called because their grains were gathered (*leguntur*)), also enrich the soil and should 'be planted not so much for the immediate return as with a view to the year later'. He goes on to say that it was customary to plough under broad beans and lupins before the pods were formed, as green manures in place of dung, if the soil was light.

By the following century Columella had added peas, chickpeas, lentils and other legumes to the list, and the practice of rotational green manuring for fertility maintenance seems well established. Columella and Pliny both list several schemes of crop rotation involving not only cereals and legumes, but other crops such as turnips for winter feed. Nevertheless, historians have been at odds as to how extensive the Romans' use of green manuring for fertility maintenance actually was. It became fashionable to deny their widespread use on the grounds that the Roman writers on agriculture were men of leisure, theoretical agronomists and counsellors of perfection. This is not the impression Columella gives when describing how to yoke and work an ox. K. D. White[221] has argued that Virgil's detailed account in the *Georgics* of alternatives to fallowing implies their widespread use.

Moreover, the Romans were not alone in this practice. At about the same time as Varro and Columella were writing in Italy, the Chinese were recording several parallel advances in agriculture[22]. Green manuring (*lü fei* = green fertilizer) may have begun with the cutting down and ploughing in of weeds in late Shang times, three millennia ago. The earliest Chinese reference to the cultivation of green manures is in the *Chhi Min Yao Shu*, which advises the ploughing in of adzuki beans in a passage apparently dating back to the first century BC. Because the maintenance of soil fertility was more dependent on the supply of animal manures in early Europe than in China, green manuring became a more important component of fertility maintenance in China, and fallowing was regarded as a last resort there. Continuous cropping with a variety of rotations was well established in many parts of China by the first century BC, thanks to the eclectic array of manures used by Chinese farmers, not only green but also animal, human, composts, ashes, bones, silkworm droppings, hair, etc.

For the next 1500 years in Europe, soil fertility and the supply of animal manures severely limited cereal yields with the result that increases in population had to be matched by expansion of the area of arable land. However, the marling and crop rotations devised by the Romans and Chinese at least helped to prevent long-term declines in cereal yields, if not their disastrous variation from year to year.

## 4.11 European agriculture in the Middle Ages

The Middle Ages were by no means a millennium of standing still in European agriculture. Jean Gimpel refers to them in his book *The Medieval Machine* as 'one of the great inventive eras of mankind', especially in harnessing new sources of power, such as the horse for agriculture and flowing water or wind for flour mills. Building on the heritage of Roman skills and organizing capacity, several innovations interacted to transform not only agriculture but also ways of life in the country. If there was a period of agricultural stasis in Western Europe, it was more apparent from the 14th to the 16th centuries, except in the Low Countries and Britain, than in the preceding millennium. For most of Europe, yields of cereals at the beginning of the 14th century were not surpassed until the 18th or, in parts of France, the 19th century.

Northern Europe in the 5th century AD was covered by forests in which oxen were grazed, while cleared patches under temporary cultivation grew mostly wheat, barley and rye in the Roman two-course rotation of one year in autumn-sown crop followed by one in fallow. As the millennium progressed and the landscape became more settled, the crops became more diversified. Oats, known to Virgil as a nuisance, became a major cereal, and spelt wheat, buckwheat and several legumes became more significant in diets with migration northwards.

Associated with the change in crops was the shift from the two-course to a three-course rotation, with the first year in autumn-sown winter wheat or rye, the second in spring-sown oats, barley, peas,

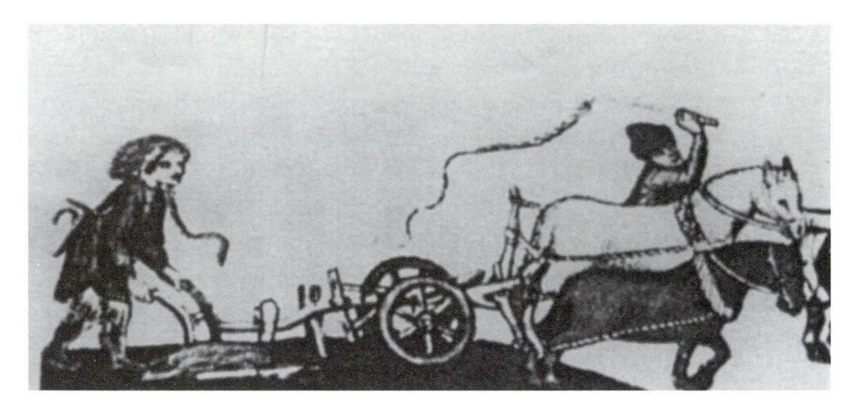

**Figure 14** A team of horses with rigid collars yoked to a high-wheeled plough. (Courtesy of the Vatican Library, Rome.)

beans, lentils or chickpeas, and the third under fallow. Romans such as Columella had considered this possibility. However, it was better suited to more northern climes and was strongly promoted by Charlemagne around 770 AD. It has been called 'the great agricultural novelty of the Middle Ages in Western Europe'.[151] However, it spread slowly because it took time to achieve the needed rearrangements of both fields and religious calendars. The advantages of the three-course rotation included: increased crop production; less risk of a famine; more evenly-spaced use of labour and draught power; better diet and improved soil fertility from the legumes and more abundant production of oats as feed for horses, at least in England if not in Scotland according to Dr Samuel Johnson.

The plough also evolved during the millennium. The wheeled plough, permitting use of a heavier share, was introduced early on, the heavy plough by the 6th century, and the mouldboard by the 11th (Figure 14). The French historian Marc Bloch may have attributed too much influence on medieval life to the plough, but its improvement had a very wide impact. More effective ploughing meant that the cross ploughing needed with the scratch plough became unnecessary, reducing labour. Fields therefore no longer had to be square in shape, and cultivation in strips, with better drainage and raised beds, became possible. Cropping was no longer confined to lighter upland soils, and heavy valley bottom soils could be cultivated. Deeper ploughing also increased root zone access to nutrients.

The fourth major change was the shift to horse power, made possible by the more frequent cropping with oats in the three-course rotation. A peasant with 8 hectares could afford to keep the two oxen required for a scratch plough by grazing them on fallow land and in the woods. They did not need shoeing, and eventually could be sold for meat. But they were slow, and eight or more of them were needed to pull the heavier ploughs. Horseshoes began to be used in Europe by the 11th century, and stronger breeds of horses were introduced both for road transport and to carry heavily-armoured knights, but their use in ploughing was not extensive until the padded horse collar was introduced from China early in the 10th century.

Widespread ploughing with horses in the 11th century led to a need to consolidate and rearrange peasant holdings. Moreover, the peasant no longer had to live so close to his land, so there was a shift in the population from small hamlets to larger villages, with more educational and commercial opportunities, and more cooperation between peasants. Especially in England, where the manorial system did not develop until the 11th century, the landlord–tenant nexus became more flexible.

Although the peasant holdings far exceeded the lord's demesne in total agricultural production, nearly all our knowledge of crop yields comes from the records of manors and monasteries, which have been intensively analysed for several European countries. Taken as a whole yields remained low and unchanged throughout the Middle Ages. Average wheat yields in England, for example, were still only 0.8 tonnes per hectare in 1450 AD[197]. Famines were frequent, that of 1314–16 being widespread through western Europe. What with the plague as well, the 14th century was not a good time to be alive, as Barbara Tuchman puts it. It has been suggested that crop yields were limited by low market demands, but it is likely that low soil fertility, particularly low nitrogen status, was the major limitation[123]. Animal manures and ley pastures, supplemented to a small extent by rainfall, provided a relatively stable but low input of nitrogen to the soil, and it was not until the end of the 15th century that further innovations made much higher yields possible.

CHAPTER 5

# Towards the first billion (1500–1825)

## 5.1 Introduction: from subsistence to commercial farming

Although there had been several earlier bursts of population growth, e.g. following the Neolithic Revolution and in early medieval times, sustained rapid growth really began in the 16th century, particularly in Europe and China.

Late medieval agriculture in Europe was still largely for the subsistence of the predominantly rural population, within a manorial framework, and with a strong focus on the cereals of Near Eastern origin. But over the three centuries following the discovery of the New World by Columbus, farming in Europe became more commercial and based on a wider range of crops used more systematically in rotations designed to improve soil fertility and to minimize the problem of weeds.

In Europe these changes began in the 14th century among the small farmers of the Low Countries, where they became full owners of their land somewhat earlier than those of France and the countries of eastern Europe. Freed from the constraints of communal ownership and rotation, the more enterprising began to replace the fallow in the universal three-course rotation with turnips for winter feed, thereby increasing both their carrying capacity for livestock and the supply of manure for their cereal crops, while still cultivating for weed control. This led on to 'convertible husbandry' in which the three-course rotation was followed by a ley pasture for several years. These leys were then improved by the sowing of clover and 'artificial' grasses, i.e. species like ryegrass not prominent in the natural grasslands. Gradually, also, as surplus cereals were imported from the Baltic, the Low Country farmers began growing the more profitable 'industrial' crops, such as hemp and flax for weaving, woad and madder for dyeing, hops for brewing and coleseed for oil. New World crops such as tobacco and potato were also grown following the voyages of Columbus and his successors.

The breakdown of feudalism began somewhat later in England than in the Low Countries but enclosure of the commons, the key to agricultural improvements, was intiated there in the 16th century and proceeded throughout the 17th and 18th centuries. The population of England trebled over this period, road and canal systems were greatly extended, industry burgeoned and cities multiplied. On the one hand, therefore, there was a growing demand for the 'export' of food from the countryside to the cities while on the other there was an accumulation of capital to be invested in country estates and their improvement.

One manifestation of this interest by the gentry was the succession of books on new farming practice. Thomas Tusser's *Five Hundred Good Points of Good Husbandrie* (1573) pressed the advantages of 'severall' (i.e. enclosed) versus 'champion' (i.e. open-field) tenure. Sir Richard Weston gave the first description of the new practices in his *Discours on the Husbandry used in Brabant and Flanders* (1645). Jethro Tull argued the case for drilling crops in rows in his *Horse-Hoeing Husbandry* (1733). There were also many agricultural surveys, like Arthur Young's tours in the various regions of England (1768–1771), Ireland and France, and his *Annals of Agriculture* which extended from 1784 to 1815. But although the writings and example of the gentry were important, particularly in the synthesis of the various improvements into a coherent, synergistic and sustainable system of farming, epitomized by the Norfolk four-course rotation, the real pioneers were often small farmers, as they had been in the Low Countries. Scientists were interested in these agricultural innovations, but at least for the years 1660–1780, scientific institutions had little effect on agricultural progress. Instead, 'the arrow of influence' led from agriculture to progress in science[117], but that was soon to change.

Although much of the increase in agricultural production to meet the doubling of world population came from extension of the area of arable land, new crops, more frequent cropping and higher yields also played an important part. Of the many New World crops introduced following the voyages of Columbus, the potato was most prominent in northern

Europe and maize in southern Europe[111], while maize, potato, sweet potato and peanut soon transformed upland agriculture in China[97]. As Paul Mangelsdorf put it, maize spread the fastest, along with tobacco and syphilis. Intensification of cropping came from replacement of that third of the arable in fallow by cultivable and highly productive crops such as turnips and potatoes. With the improved system of soil fertility maintenance, average cereal yields had increased by about 50% by the end of the 18th century in England and the Low Countries but elsewhere in Europe by no more than 10%[78]. Between them, these improvements had almost doubled food production in England.

More than that, however, these increases were accompanied by a doubling of the productivity of farm labour in England as improvements in management became widely adopted between 1700 and 1850[150,227]. More labour was freed for industrial development, and so the synergistic interactions between agricultural improvement, industrial development, population growth and the rising demand for food continued, with no clear indication of a prime mover.

It may seem ironic, therefore, that The Reverend Thomas Robert Malthus (1766–1834) published his pessimistic concerns about population towards the end of this era of progress and optimism. In fact, it was the optimistic excesses of Condorcet and William Godwin that provoked him to write his essay raising the issues which revisit us at each step in this history of the relations between population and agriculture. Malthus' concerns were shared on the other side of the world, where the population of China more than doubled during the 18th century. 'The Chinese Malthus' Hung Liang-chi (1746–1809) wrote two essays on the population problem in 1793 in which he also argued that increase in population could outstrip the means of subsistence, bringing misery, sickness and famine to many because 'the government could not prevent the people from multiplying themselves'[97]. Indeed, until the beginning of the 19th century, despite the global population being less than one billion, the majority of the people in the world, including those in Europe and China, were probably in a chronic state of hunger[42,78].

However, at least the frequency of famines in western Europe decreased through the 17th and 18th centuries, the last being in the 1620s in England, 1732 in Germany and Scandinavia, 1795 in France, and 1845 in western Europe as a whole (in Ireland). Although the Irish famine occurred about 20 years after the world population reached one billion, its seeds were sown and its likely occurrence predicted by several writers at the end of the 18th century. Often viewed as an example of the Malthusian scenario, recent research suggests that it was not demographically inevitable, and that Ireland was more under-industrialized than over-populated.

## 5.2 The impact of Columbus

Thomas Jefferson once wrote: 'The greatest service which can be rendered any country is to add an useful plant to its culture … .' What is more, he practised what he preached, enriching his country's agriculture by smuggling rice seeds out of the Italian Piedmont, risking execution by the Piedmontese. Plant introduction has, fortunately, not always been so hazardous. Seafarers both distributed and collected seeds habitually on their voyages of discovery. On November 5, 1492 two of Christopher Columbus' crew returned from a trip to the interior of Cuba with 'a sort of grain they called maiz which was well tasted, bak'd, dry'd and made into flour'. Thus began the introduction of plants from the New World to the Old, and of introductions in the opposite direction the following year with sugar cane and citrus.

Long before that, however, domesticated plants had been taken to or collected from new regions. The earliest recorded collecting expedition is that of Queen Hatshepsut to Punt, on the southern shores of the Red Sea, about 3.5 K years ago, but that was preceded by the demic diffusion of plants and people westwards across Europe, eastwards to India and south to Ethiopia from the Fertile Crescent; by the transfer of sorghum and finger millet from East Africa to India (and of *durra* sorghum back to Africa in Islamic times); and by the westwards movement of sugar cane and bananas from New Guinea and south-east Asia.

Such early movements of crops are reflected in the 'Old English' origins given for their names in the Oxford English Dictionary, e.g. for wheat, barley, oats, rye, peas, beans (*Vicia*), linen and hemp. Then came a wave of crops with names of Middle English origin, such as millet (*Setaria*) from north China, cabbage and olive from the Mediterranean, and many plants brought to southern Europe following the expansion of Islam in the 8th century, such as the date, orange, lemon, fig and lime. Along with these plants the Arabs brought their experience with irrigation in Mesopotamia and Egypt, and with the intensive cultivation of fruits, vegetables and ornamental plants. Although their fostering of horticulture was important to Spain, they also introduced sugar cane, cotton, rice, the mulberry and lucerne into southern Europe[77].

However, the two-way redistribution of crops in the years following the voyages of Columbus dwarfs all others in its impact on world food production. About one quarter of all the plants that have been domesticated come from the Americas and their names were soon recorded in the English language: the kidney bean (1548), cassava (1555, = manioc 1568), maize (1565), potato (at first the sweet potato (1565) and then the true potato (1597)), tobacco (1577), tomato (1604), and pineapple (1664), first tasted by Columbus on 4 November 1493.

The sweet potato (*batata*) was presented to Queen Isabella by Columbus after his first expedition and by 1521 had spread to the Far East via Africa[28]. By contrast, there is no record of when the true potato (*patata*) was brought to Europe after 1537, when the Spaniards first encountered it in Colombia (Figure 15). Early Spanish records clearly distinguished *batata* from *patata* but most early English records did not. The first published account of the potato was in 1552, and it was probably brought to Europe before 1562 because it was already being exported in quantity from the Canary Islands by 1567[89]. Many other American domesticates were likewise rapidly adopted and extensively grown not only in Europe but also in Africa and Asia, thanks to the Portuguese seafarers. In Europe the growing of maize was at first strongly discouraged by the Church because it was not included in the list of crops to be tithed.

One reason for the quick success of the American crop plants, the majority of which came from high altitudes in Mexico and the Andes, was their better adaptation to cooler growing conditions than that of many of the crops from Asia and Africa. With one small exception, all Vavilov's centres of origin/diversity lie between latitudes 38°N and 15°S, mostly at low altitudes. Tolerance of cool conditions is not quickly acquired, as both Alphonse de Candolle and Charles Darwin recognized, so the American domesticates provided particularly valuable alternative crops for the higher latitudes.

Also operating in their favour was the fact that crops taken to distant new environments often leave many of their pests and diseases behind and may therefore perform better than in their original habitat. Their pests and diseases may eventually catch up with them, with disastrous

**Figure 15** Digging with a foot plough and planting potatoes in the Andes at the time of the Spanish Conquest. From a 16/17th century manuscript by Felipe Guaman Poma de Ayala (with permission of the Institut d' Ethnologie, Paris).

consequences as we shall see, but even today, for most crops, their average yield beyond their centres of origin is substantially higher than that within it.

Undoubtedly, therefore, the spread of American domesticates through the Old World in the wake of Columbus led to an increase, as well as a diversification, of the world's food supply. In China the population had hovered around 60 M for several centuries but increased to 150 M by 1640 AD. Ping-ti Ho[97] attributes this partly to the spread of much earlier-maturing Champa rice varieties, permitting greater cropping frequencies, and partly to the introduction of New World crops adapted to upland conditions, such as sweet potato, maize and peanuts which were being grown in China within 50 years of Columbus's discovery of the New World.

## 5.3 The potato in Europe

Although it originated and was domesticated in South America, and was introduced to Ireland, unheralded, only late in the 16th century, the potato was already known as the 'Irish potato' by the middle of the following century. The so-called 'potato famine' of Ireland was still two centuries in the future.

The genus *Solanum* includes more than 230 species ranging in habitat from the Peruvian and Chilean coasts to the high Andes. The potato has more wild relatives than any other crop, encompassing a wide range of tolerances which have been drawn on in the course of its evolution. Like wheat it is a polyploid crop with domesticated species at several levels of ploidy from two- to five-fold. Its original domestication probably took place in the region of Lake Titicaca, between Bolivia and Peru.

There is archaeological evidence of wild potatoes being gathered 13 K years ago in southern Chile and 10 K years ago in several Andean caves. The earliest evidence of domestication comes from Chilca Canyon in Peru about 7 K years ago. The diploid species *Solanum stenotomum* is probably the original domesticate, but three wild species have been involved in the further evolution of the potato since its domestication. Two of these led to the development of frost resistant potatoes, the other to the tetraploid potato *S. tuberosum*, subspecies (ssp.) *andigena*. It was this subspecies, not our modern potato, that was so important to

Inca culture and whose tubers were central to their ceramic art. These were the *papas* first encountered by the Spanish in 1537, described by Juan de Castellanos as 'white and purple and yellow, floury roots of good flavour, a delicacy to the Indians and a dainty dish even for the Spaniards.'

Coming as they did from high altitudes but low latitudes, *andigena* potatoes were well adapted to the cool conditions but not to the long days of European summers. Although it was not understood at the time, the long days of summer prevented tuber formation until close to the autumn equinox, leaving little time for tuber growth. The early botanical descriptions of the potato in Europe, and its initial problems with late tuberization, tend to support the conclusion by both Redcliffe Salaman – who devoted his whole research career to this crop and produced a wonderful testament to his obsession in *The History and Social Influence of the Potato*[181] – and J.G. Hawkes[88] that ssp. *andigena* was the foundation of the modern potato.

On this assumption both the response to daylength and the growth habit must have been progressively shifted by selection among seedling variants of *andigena* towards those of ssp. *tuberosum*. This could account for the slow adoption of the potato in many European countries, where it was not widely grown until the middle of the 18th century. In Ireland, however, where both the climate and soils were favourable, the potato became what Salaman describes as 'the universal and staple article of the people's food in the greater part of the island' within fifty years of its introduction in the late 16th century.

However, selection within ssp. *andigena* did not produce really early ('first early') varieties, with tubers ready for harvest two months after planting. Most of these can be traced back to some 'Rough Purple Chile' potatoes purchased in 1851 from a market in Panama by the Reverend Chauncey Goodrich. The progeny of these yielded well in New York state and a variety derived from them, Early Rose, has been involved in the breeding of most of the very early-maturing European varieties.

Described as 'The greatest gift of the New World to the Old', the potato was an excellent crop for those with very small holdings, particularly in areas where the law of primogeniture did not apply, so that even small peasant holdings were progressively subdivided. Its cultivation required only a spade or a hoe; it took only 3–4 months for the crop to mature; it could be grown on a wide variety of soils; it yielded 3–4 times more food than cereals and often flourished when cereal crops failed;

nutritionally it combined well with milk, and a single acre (0.4 hectares) of the crop, plus a cow, could feed a family for much of the year. Therein lay its popularity, and its threat.

## 5.4 High farming in the Low Countries

The Flemish farmers of the 16th and early 17th centuries were more interested in improving their husbandry than in writing about it. Fortunately Sir Richard Weston, a political refugee from England, was so impressed by it during his visit in 1644 that he wrote the first description of their farming practices in his *Discours on the Husbandry used in Brabant and Flanders: showing the wonderful improvement of land there; and serving as a pattern for our practice in this Commonwealth* (1645). Moreover, he gave the lie to G.B. Shaw's dictum that 'He who can, does. He who cannot, teaches' by successfully applying Flemish practices of drainage, irrigation, enrichment and rotation for many years on his own estates at Worplesdon in Surrey, more than a century before they became widely used in England.

As population pressure in the Low Countries built up in the 16th century to the highest in Europe, many farmers had only a small area available for cultivation, but they had full ownership and control of their land before farmers elsewhere[77]. Because the Low Countries relied to a considerable extent on relatively cheap grain imported from the Baltic, their farmers could intensify their farming practices and concentrate on more profitable crops such as the linen flax, hemp and dyes (such as woad, madder and saffron) required by the local weaving and lace industries, hops for beer-making, tobacco, cole seed and the market gardens and orchards needed to serve the growing towns. They transformed farming from a subsistence to a partly industrial activity.

Just as the inhabitants of the Low Countries became expert in the arts of canal and lock building, so also did they develop those of agricultural drainage, reclamation and survey. Their pre-eminence in these activities resulted in their skills being imported into England to drain the fens and 'levels' on a grand scale between the 15th and 17th centuries AD. Their drains were usually filled with stones, pebbles or shafts of wood, and then turfed over. It remained for an English gardener, John Reade, to invent in 1843 the cylindrical clay pipes which came to dominate agricultural drainage. In his great history of English farming, Lord

Ernle repeatedly emphasizes the need for improved drainage on both pasture and arable land.

But the Low Countries contributed more than better drainage to the intensification of agriculture. Their specialization on industrial crops, horticulture, fruit-growing and livestock farming shifted the perspective on farming profoundly, bringing an influx of capital for land reclamation and improvement. As Robert Child wrote in 1650: 'the improving of a Kingdome is better than the conquering a new one.'

Farming in the Low Countries also laid great emphasis on livestock, particularly cattle, whose manure fertilized the crops. Turnips were introduced as a winter feed crop for cattle. Then came the deliberate improvement of pastures not only by drainage but also by the widespread sowing of clovers and of what were referred to as 'artificial grasses'. For Arthur Young the 18th century usage of 'artificial' served the promotional functions that 'synthetic' served in the 19th and 'genetically engineered' in the 20th centuries, and these practices quickly spread beyond the Low Countries. Pastures were no longer merely the pickings available on fallow land, but were treated as a sown crop for livestock and for improvement of the soil, the beginnings of 'convertible husbandry.'[77] Clover seed was on sale in London by 1650. Fifty years later enough to sow two thousand hectares was imported, mostly from Flanders, and by 1800 enough to sow 200 K hectares, implying that one quarter of the arable land in England was by then under the Norfolk rotation[69].

Driven partly by population pressure and partly by emergent industrialization under conditions where alternative sources of the staple cereals were available at reasonable cost, the farmers of the Low Countries initiated the intensification of agriculture by which cereal yields there and in England rose sharply in the 17th century and were not matched elsewhere in Europe for many years. Henceforth, as Slicher van Bath[193] put it 'farmers became conscious of the market economy and farming practice was influenced by the price level for agricultural products'. Dutch farming became a Mecca for agriculturists from all over Europe.

## 5.5 The Norfolk agricultural 'revolution'

The Norfolk agricultural revolution exemplifies the complex interrelations between population growth, industrialization and agricultural development. Its roots in England go back to the beginnings of enclosure of the 'commons' in the 16th century, well before Sir Richard Weston returned from the Netherlands. Its full flowering was reached with Coke of Norfolk (1754–1842) early in the 19th century.

Indeed, unlike the earlier advances in Flanders out of which it grew, the Norfolk revolution came to be epitomized by the writers who publicized the advantages of its various components. After Weston came Jethro Tull (1674–1740) with his *Horse-Hoeing Husbandry*, 'Turnip' Townshend (1674–1738) with his enthusiasm for his eponymous crop and for marling, Robert Bakewell (1725–1795) who selected his sheep and cattle for their meat production, Arthur Young (1741–1820) with his voluminous advocacy of enclosure and experimentation, and finally Thomas Coke of Holkham, the great publicist for the Norfolk four-course rotation. These are the men to whom Lord Ernle gives most credit in his great book *English Farming Past and Present* for the 'agricultural revolution', which he locates in the 18th and early 19th centuries. Other historians, however, have placed the several innovations earlier, in the 17th century, and among the *hoi polloi* rather than the landed gentry.

Between 1700 and 1830, the English population grew from 5.1 to 13.3 M and the number of horses trebled, yet both horses and people were fed and, for a period, England was also 'the granary of Europe'[57]. Moreover, this great expansion of agricultural production was achieved along with a fall in the proportion of the population engaged in agriculture from 55% in 1700 to only 25% in 1830[150], thereby freeing labour for the new industries and creating a strong demand for cereals in the towns.

Throughout this whole period the area of arable land continued to increase, as did the frequency of cropping, but the key to the advances in agricultural production was the progressive enclosure of the commons, begun in 16th century Shropshire to turn arable into grassland but promoted under King George III for the opposite purpose, in response to relatively higher cereal prices.

Arthur Young's *General Report on Enclosures*, prepared for the Board of Agriculture in 1808, leaves the reader in no doubt that the trauma of enclosure of the commons for the many who were dispossessed was essential for the progress of English agriculture: 'Common field agriculture usually lies ... in so scattered and divided a state that every operation of tillage, harvest etc. is carried on at an expense considerably greater than in enclosures ... (etc.) ... But all these are trifling points when compared with the fatal shackles under which every occupier, however anxious for improvement, is bound to the courses of crops entailed by custom on a common field ... What a gross absurdity, to bind down in the fetters of custom ten intelligent men willing to adopt the improvements adapted to enclosures because one stupid fellow is obstinate for the practice of his grandfather! To give ignorance the power to limit knowledge ... and fix an insuperable bar ... to all that energy of improvement which has carried husbandry to perfection by means of enclosure.'

Almost 40 years earlier, in 1771, Young had referred to 'the Norfolk system' as having seven points: 'first, enclosures without assistance from Parliament; second, use of marl and clay; third, proper rotation of crops; fourth, culture of turnips, hand-hoed; fifth, culture of clover and ray (rye)-grass; sixth, long leases; and seventh, large farms.' In fact, marling was an ancient practice, well known to the Romans. Turnip crops had been grown in England since the early 17th century. They were used on less than 2% of Norfolk farms in the 1660s, but on more than half of them 60 years later. Over the same period the use of clover increased from less than 1% to 24% of Norfolk farms[149]. This was the period in which the Norfolk four-course rotation became widely established: turnips in the first year, as fodder for cattle or 'folded' for sheep; thus manured and well cultivated, the soil was then suited to a crop of wheat or barley, with the straw later used for farmyard manure; this was followed by a clover ley to restore soil nitrogen and provide grazing; finally another crop of barley or wheat was commonly taken.

Note that in his 1771 list of seven points, given above, Arthur Young specified 'turnips, hand-hoed'. This was almost 40 years after Jethro Tull had published his *Horse-Hoeing Husbandry* with his persuasive argument for sowing crops in drilled rows so that seed was saved and horse-drawn 'hoes' could replace the armies of 'sarclers' needed to keep the weeds at bay. The advantages of drilled rows, including economy in the use of seed for sowing, had been recognised by the Chinese in the 6th century AD, and usable drills had been designed by them and by others before Tull. Yet it was not until the 1790s that his labour-saving system of cultivation came into widespread use, leading to a second phase of improvement. Although many of the individual components of the English agricultural revolution had a long history, it was the synergistic interactions between them in the Norfolk system that made it such an effective agent of improvement, as Naomi Riches illustrates in *The*

*Agricultural Revolution in Norfolk* (1937). And it was the combination of agrarian with agricultural change that was so powerful.

The tragedy of the commons, to use Garrett Hardin's phrase, was unavoidable. As Lord Ernle[57] put it: 'The divorce of the peasantry from the soil, and the extinction of commoners, open-field farmers, and eventually of small freeholders, were the heavy price which the nation ultimately paid for the supply of bread and meat to its manufacturing population.' Along with these losses the farmer became less of a husbandman and more of an entrepreneur.

## 5.6 Malthus and his Essay on population

Thomas Robert Malltiss (changed to Malthus in 1792) was born in 1766, first published his *Essay on the Principle of Population* in 1798, and died in 1834. His life therefore coincided with the 'Norfolk revolution' in agriculture, and with the shift from agriculture as the dominant economic activity in England to its being overtaken by industry and urbanization.

The 1790s were a particularly turbulent decade in England with rapid population growth and industrial expansion, a series of poor harvests from 1794 to 1800, high food prices, and widespread concern and several government enquiries about the vulnerability of the nation's food supply. Popular unrest reached a peak in 1795. This was the background against which Robert Malthus wrote his first Essay[209]. It was an opportunity for a general treatment and he seized it.

The young Malthus began his *Essay* boldly with his famous statement on the relation between rates of increase of population and of food supply. With Benjamin Franklin's observation on the prolificacy of plants and animals in mind, and after considering the demographic evidence, particularly from America, he concludes: 'It may safely be pronounced therefore that population when unchecked goes on doubling itself every twenty-five years, or increases in a geometrical ratio.' As for the food supply he thought a doubling in Britain in the first 25 years would be 'a greater increase than could with reason be expected', and likewise 'every 25 years by a quantity equal to what it at present produces: the most enthusiastic speculator cannot suppose a greater increase than this.' Therefore 'the means of subsistence, under circumstances the most favourable to human industry, could not possibly be made to increase faster than in an arithmetical ratio.' Thus, the further increase in world population, which he correctly estimated to be about one billion at the time, 'can only be kept down to the level of the means of subsistence by the constant operation of the strong law of necessity acting as a check upon the greater power.'[129]

This elementary but emotively powerful mathematical contrast, delivered in trenchant prose with polemical purpose, had an immediate political impact. Yet Malthus must have known he was on shaky ground for increases in the means of subsistence, as suggested by his more cautious wording in his *Summary View of the Principle of Population*[130]. Malthus was writing his *Essay* at a time when the English arable was being extended and yields were rising rapidly under the impact of the Norfolk rotation. Comprehensive but unpublished surveys of the yield of wheat in England from 1809 to 1859 suggest that it may indeed have doubled within the 25 year period between 1835 and 1859[91]. When combined with the increase in arable area, this would have increased production by more than Malthus' most enthusiastic speculator could suppose.

In 1798 there was no international or even national data base by which to estimate increases in food production but there was also no strong reason why Malthus should have summarily excluded the possibility of a geometric increase in food supply. It is true that Adam Smith and David Ricardo also rejected the possibility of geometric growth in the economy, because of diminishing returns on a limited supply of arable land, but I suspect the young Malthus also recognized the dialectical force of the geometric/arithmetic contrast. It has been said of Malthus that no other social scientist has been both attacked and defended with so little regard for what he actually wrote. Part of the problem has been the extensive changes he made in the course of the six editions of his *Essay* as he acquired more data and widened his experience.

Of the first edition of the *Essay*, Malthus said: 'It was written on the spur of the occasion and from the few materials which were within my reach in a country situation.' J.M. Keynes thought it displayed 'the brilliance and high spirits of a young man ...', whereas in the second edition (1803) 'political philosophy gives way to political economy, general principles are overlaid by the inductive verification of a pioneer in sociological history.' Malthus first admitted in his second edition that other 'preventive' checks to increase in population besides his 'positive' check could be operative, leading him to soften some of the harshest conclusions of the first *Essay*. His views on the Poor Laws brought him into conflict with many, including Arthur Young, and with the many economists of the time who considered that increase in population was desirable.

In relation to neither industrialization nor agriculture, let alone military capacity, did Malthus deny this: 'That an increase of population, when it follows in its natural order, is both a positive good in itself, and absolutely necessary to a further increase in the annual produce of the land and labour of any country, I should be the last to deny. The only question is, what is the natural order of its progress? In this point Sir James Steuart ... determines that multiplication is the efficient cause of agriculture, and not agriculture of multiplication.'[129] Steuart's view is, of course, that subsequently proposed by Boserup, which Malthus goes on to discount.

In the fifth edition of 1817, referring to the crucial question of the 'natural order of progress', Malthus added: 'In this progress nothing is more usual than for the population to increase at certain periods faster than food; indeed it is part of the general principle that it should do so; and when the money wages of labour are prevented from falling by the employment of the increasing population in manufactures, the rise in the price of corn which the increased competition for it occasions is practically the most natural and frequent stimulus to agriculture. But then it must be recollected that the greater relative increase of population absolutely implies a previous increase of food at some time or other greater than the lowest wants of the people. Without this, the population could not possibly have gone forward.'[129] Thus, people, prices and production, both agricultural and industrial, are indissolubly linked, and the question of prime mover becomes a philosophical one.

## 5.7 The Irish potato famine

The potato famine of 1845/46 looms over Irish history but also over demography through having been viewed as a timely example and justification of the concerns expressed by Malthus. Indeed Malthus and others had forewarned of the likelihood of a food supply crisis in Ireland at various times during the fifty years before it actually occurred. The main reasons for their concern were the rapid increase in population – said by Malthus to be the fastest outside America – and the almost total dependence of many of the poorest people on one food crop.

Given the overwhelming importance of the potato in Ireland by the end of the 18th century, it is surprising how little is known about its early adoption and spread there. But it was well adapted to the more humid areas and it complemented the cereals in the sense that poor years for cereals tended to be good years for the potato and vice versa. In a nutritional sense the potato also complemented the milk products which were a significant component of Irish diets. Even Arthur Young had to admit that the potato must be a good food. Referring to the Irish country folk he wrote 'When I see (their) well formed bodies ... their men athletic and their women beautiful, I know not how to believe them subsisting on an unwholesome food.'

Between Young's visits in the 1770s and the onset of the famine, the population of Ireland doubled to reach 8 M, for which the potato is often held to blame. The relative yields of potato and cereal crops at that time were such that throughout northern Europe, 3–4 times more people could be fed by an acre of potatoes than by an acre of wheat. Today, by contrast, average wheat yields in Ireland are over 40% higher than those of potatoes in terms of edible dry matter. Crop yields in late 18th century Ireland were as high as those in England and much higher than those in France[148]. Quite small holdings could provide all the food required by a family, but little else. Because most landowners did not require the labour of their tenants, rents were paid by the sale of cereals or, in earlier years, of household crafts such as linen.

In his first *Essay* Malthus had written: 'if potatoes were to become the favourite vegetable food of the common people ... the country would be able to support a much greater population, and would consequently in a very short time have it.' Nevertheless, on the eve of the famine potatoes accounted for rather less than a quarter of the value of overall Irish agricultural output compared with almost 40% for other crops and 22% for cattle and dairy products[71]. Given the extent of agricultural exports from Ireland, even a halving of the potato crop in 1845 and 1846 need not have created a famine in which 800,000 *additional* deaths occurred as well as accelerated emigration. The excess mortality was far higher than in more industrialized countries such as Belgium, where the potato crop also failed. The problem was that although potatoes made up less than a quarter of Ireland's agricultural production, they were nevertheless the predominant staple food of more than 3 M people as well as a major element in the diet of the remainder, and if their landlords demanded their cereals for rent, starvation was inevitable.

Many earlier writers have viewed the Irish famine as demographically inevitable, but recent research suggests this was not so. In his book *Why Ireland Starved*[139], Joel Mokyr shows that although the Irish cherished children and had high marital fertility rates, age at marriage was not unusually young, and that the population growth rate was slowing after 1790. He estimates that the diet of even the small farmers was more than adequate, as Arthur Young had implied, and that there was more arable land per person in Ireland than in Belgium or England. Mokyr concludes that Ireland was not overpopulated, and not poor – and therefore vulnerable – because of its demography but for other reasons, especially its lack of industrialization.

That Ireland would be undone by a failure of the potato crop had been anticipated by some but that the cause was a fungal parasite was neither foreseen nor believed. After all, Louis Pasteur's germ theory of disease was still 25 years away in the future.

Late blight of potatoes, long known in Mexico, may have made its first appearance in Europe, at Liège, in 1842, and two years later at Lille. In August of 1845, after a period of cool, wet weather, Belgian potato fields were reported to be devastated by a 'murrain' far more destructive than scab or curl. This spread rapidly (Figure 16) and by early September, from Poland to England, potatoes of all varieties had succumbed. John Lindley, the editor of *The Gardeners Chronicle and Agricultural Gazette,* had sounded the alarm. Then on September 13 he announced: 'We stop the Press, with very great regret, to announce that the Potato Murrain has unequivocally declared itself in Ireland ... for where will Ireland be, in the event of universal potato rot?'

The rest is history, nearly all of it with emphasis on the causes – demographic, geographic, economic, political or social – of the famine. Yet the immediate cause of the disaster, the murrain, had several major effects on subsequent agricultural history. For one thing, with the repeal of the Corn Laws in 1846, it turned the United Kingdom away from a policy of self-sufficiency and protection of domestic agriculture towards free trade and inter-dependence. And it led to the first recognition that 'murrains' and diseases of crops could be caused by microorganisms and might therefore be controllable.

The redoubtable Dr Lindley proposed that the murrain, apparently

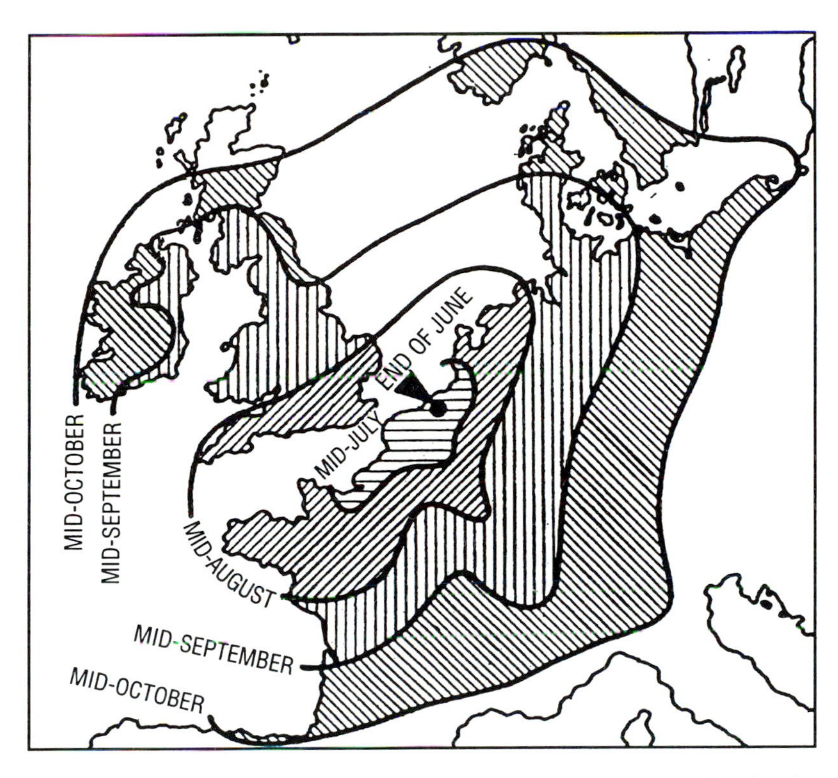


**Figure 16** The spread of potato blight across Europe and to Ireland in 1845 (Bourke[21]).

brought on by the cool, wet weather, was due to a kind of waterlogging or dropsy of the plants. However, the Reverend M.J. Berkeley, expert on fungi, put forward the revolutionary hypothesis that the mould always associated with the murrain was its cause and not its consequence. His Parisian colleague Dr Montagne confirmed that the same fungus was always associated with the murrain in France and named it *Botrytis* (now *Phytophthora*) *infestans*[112]. But a full proof that it caused the disease eluded them for some time.

So what has been called the greatest gift of the New World to the Old was eventually joined by its greatest scourge to give rise not only to 'the last great natural disaster in Europe' as Mokyr refers to it, but also to the science of plant protection.

CHAPTER 6

# The second billion (1825–1927)

## 6.1 Introduction: the entry of science

The population of the world doubled, from one to two billion, in the century following 1825, with most of the increase still being in Asia and Europe. The period was characterized by a rise in food supply per head from less than 2000 to more than 3000 Calories per day in the countries of western Europe and, from the middle of the 19th century, by a sharp rise in the proportion of animal protein in the diet[42,78]. These changes were associated with the beginnings of agricultural research, stimulated by Humphry Davy's lectures on agricultural chemistry, resoundingly launched by Justus von Liebig from Giessen in 1840, and epitomized by the motto of the Royal Agricultural Society of England at its establishment in 1838: 'Practice with Science'.

Liebig quickly discovered the elusiveness of this desirable union, and it was many years before agricultural research began to have a major impact on the human carrying capacity of the earth. In fact, the doubling of the world's population in the century before 1927 was probably associated with a doubling of the area of land cleared of forests and prairies for agriculture. I say probably because the many estimates of earlier populations are matched by only a few uncertain estimates of arable area (Figure 17). Between 1870 and 1920, while the world population increased by 40% the arable area increased by 75% due to extensive land clearing, particularly in North America and Russia. The burgeoning populations of those areas were fed by the extension of agriculture rather than by its intensification, while in Europe there was a marked reduction in the proportion of arable land left fallow.

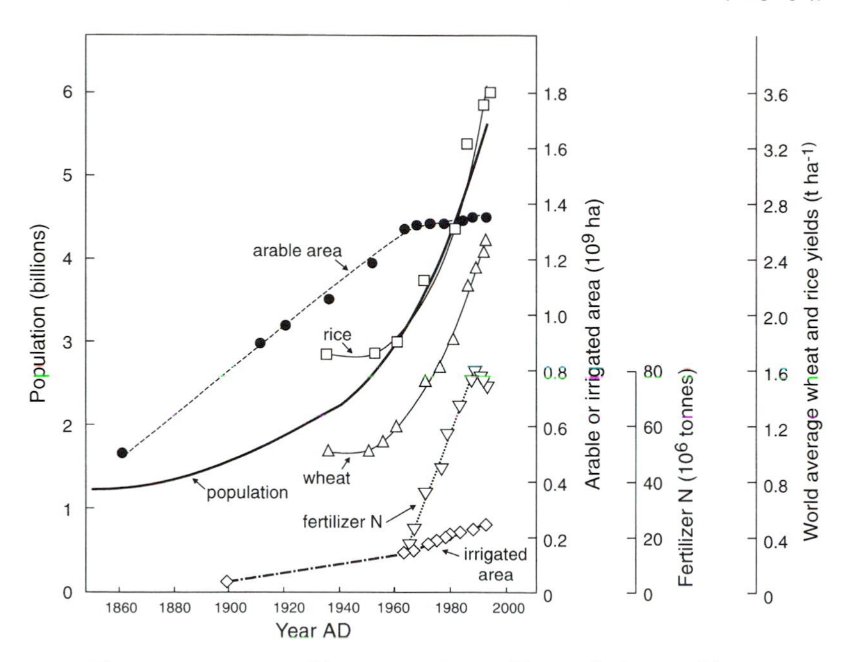


**Figure 17** Increases this century in world population, arable area, the average yields of wheat and rice, the amount of N fertilizer used, and the irrigated area of the world[59].

For the period under consideration, significant improvement in crop yields was largely confined to western Europe and Japan. Referring to England in the latter part of the 19th century, Lord Ernle wrote: 'The age of farming by extension of area was ended; that of farming by extension of capital had begun.' Average wheat yields in England increased by at least 40% between 1825 and 1927, partly from the continuing impact of the Norfolk rotation and later from the introduction of fertilizers such as Chilean nitrate, Peruvian guano, German potash and John Bennet Lawes' superphosphate. In Japan after the Meiji restoration in 1868, rice yields had almost doubled by 1927 following the adoption of several European advances, whereas rice yields in China reveal only a modest increase over the period[76]. Over-population in the lower Yangzi region led to one of the world's greatest civil wars, the Taiping rebellion from 1851 to 1865, which greatly reduced the population and persuaded many to move to other regions of China.

Crop yields on some of the cleared lands of the New World began to decline after the initially high soil fertility was exhausted. In both North

America and Australasia it was not until the 1930s that yields began to rise. Until the middle of the 19th century, crop yields were thought to be limited by the supply of farmyard manure: the more livestock the farmer could carry, the higher his yields. The discovery of nitrogen fixation in legumes by Hellriegel and Wilfarth in 1886 explained their long-recognized beneficial effects in crop rotations. Even before that, however, one-fifth of the arable land in northern Europe was under leguminous crops, which are thought to have provided more nitrogen to cereals than did the farmyard manure[41].

In the second half of the 19th century, however, the limitations on crop yields by the supplies of farmyard and green manures were lessened as mineral fertilizers began to be used, especially after the introduction of superphosphate and the demonstrations at Rothamsted of the returns in yield from the application of N, P and K fertilizers. Sir John Russell[179] referred to these fertilizers as 'the first and for long the only application of science to agriculture.'

That other ancient but periodic and unpredictable limitation on yields, by the diseases and pests of crops, also seemed on the threshold of control as a result of scientific research following the work of Millardet and others in France on the use of verdigris, Paris green and other disease and pest-controlling chemicals. Then, as our two billion approached, came the beginnings of biological control.

The introduction of new crops, e.g. of soybeans into the USA, continued throughout the 1825–1927 period, with botanical gardens playing a major role, particularly in the redistribution of crops adapted to the tropics. Aiding such redistribution was the recognition that plant responses to daylength, as well as to temperature, played an important role in the adaptation of crops to new environments. The selection of superior lines of crops was practised by many farmers and, increasingly, by seed companies such as that of Vilmorin in France, founded in 1727. This period also saw the development of plant hybridization, leading to the publication of Mendel's laws governing genetic segregation in 1866, their rediscovery in 1900 and their rapid application to plant breeding. In this, as in several other fields, empirical advances had preceded scientific understanding but the latter resulted in faster progress and greater efficiency.

As science increasingly permeated agriculture, the questions of how to support research, extend its findings and educate agriculturists in their adoption became a matter of urgency. Early in the 19th century much research was stimulated by the award of substantial prizes for specific objectives, some by governments, others by local agricultural societies. Pasteur won several of these. Another was won by Mathieu Tillet for the control of bunt (*Tilletia*) in wheat. In England J.B. Lawes continued the tradition of the landed gentry in supporting agricultural experimentation, but on a scale unmatched by his predecessors when he established Rothamsted as the first agricultural experiment station.

At the same time in Germany Justus von Liebig, with his challenging views on plant nutrition, was attracting many students from overseas to his laboratory at Giessen. The impact of his writings and those of Albrecht Thaer was such that several German states established agricultural experiment stations. The first was at Mockern in 1852, ten years after the first of the classic long-term experiments was initiated at Rothamsted and a year earlier than the establishment of an experimental and model farm at East Lansing, Michigan. The institutionalization of government funding for agricultural research in the USA began in 1862 with the establishment of the US Department of Agriculture and the provision of land to the various states to establish colleges of 'agriculture and the mechanic arts'. Similar colleges had existed in England since the first was founded at Cirencester in 1845.

The wave of public support for agricultural research and education then rolled on to Japan in the Meiji Period (1868–1911). In seeking to shed the isolation and feudalism of the preceding Shogunate the Meiji leaders first sought to mechanize Japanese farming, by analogy with changes to industry. This failing, they then brought in German agricultural chemists to remould Japanese agricultural research and education in the Liebig tradition. A Ministry of Agriculture and Commerce was established with itinerant instructors and a network of national and prefectural agricultural experiment stations.

Thus, by the time the population of the world reached two billion in about 1927, the need for publicly-funded research and education in agriculture was widely accepted even though its benefits in such areas as the nutrition, pathology, introduction and improvement of plants had only just begun to flow. Scientists even of the stature of Liebig were soon forced to recognize the complexities of agricultural systems and the need for validation in field experiments *before* giving advice. But research was the key to the material and intellectual inputs that broke through the limitations on productivity in traditional agriculture.

## 6.2 Justus von Liebig and plant nutrition

In 1840 the celebrated German chemist Justus von Liebig (1803–1873) dropped his bombshell, a book entitled *Organic Chemistry in its Application to Agriculture and Physiology*. It arose out of an invitation to address the British Association for the Advancement of Science on the state of organic chemistry. Only 34 years old and not engaged by the conference, he had spent a day at the beach with a request for publication on his mind. His choice of agriculture, in which he had read widely, was a complete surprise. But his incisive approach, combative style and scientific eminence gave his book a widespread and long lasting impact and epitomized both the strengths and the hazards of agricultural science from its outset.

In 1840, our knowledge of plant nutrition and growth was still hopelessly confused. The Aristotelian concept that plants ingested preformed organic matter from soil, hence the great value of animal manures, was still alive and well. Jethro Tull had called the soil 'the pasture of the plant', and Albrecht Thaer, in the early 1800s, had called the 'essential' organic nutrient humus. This was the focus of Liebig's invective. He pointed out that the earlier experiments of Priestley, Ingen-housz and de Saussure had shown that most of the carbon, hydrogen and oxygen in plants came from the atmosphere, via the process of photosynthesis, and not from the soil. Humphry Davy, in his earlier and influential book *Elements of Agricultural Chemistry* (1813) had also made this point yet remained ambivalent by referring to soil organic matter as the true food of roots. Liebig's trenchant review – 'Liebig at his worst' in Margaret Rossiter's words[177] – finally put paid to the humus nutrition theory.

On the positive side Liebig focused attention – particularly in the later editions of his book – on the mineral components of plants. Van Helmont's experiment in the 1640s had shown that plant growth entailed only slight loss of soil weight, and that mineral ash was only a small constituent of plant dry weight, so it tended to be neglected until Liebig brought it into sharp focus. Even Lavoisier thought the alkali metals were produced during plant growth but Davy considered it 'a much more probable idea, that they are actually a part of the true food of plants'. Thanks to improvements in analytical methods since Davy's time, Liebig could add phosphorus to his list of essential nutrients, and went on to explain the fertilizing action of animal manures in a new way. In his later editions particularly he emphasized the value of soil and ash analysis with the objective of adding back to the soil the minerals removed by each crop, a recommendation that went awry in the subsequent craze for soil analyses in the USA.

Where Liebig went wrong, and increasingly so, was in relation to nitrogen. In his first edition he had stressed that supplying crops with nitrogen was 'the most important object of agriculture'. He had estimated that the amount of ammonia in rainwater was enough for forests but not sufficient for the purposes of agriculture: 'Agriculture differs essentially from the cultivation of forests inasmuch as its principal object consists in the production of nitrogen in any form capable of assimilation whilst the object of forest culture is confined principally to the production of carbon.' He regarded atmospheric ammonia as the source of nitrogen in plants, overlooking the nitrates which Boussingault subsequently showed to be more important.

In later editions Liebig shifted his emphasis away from ammoniacal manures to the indispensability of the mineral nutrients. For no apparent reason he increased his estimate of the amount of ammonia in the atmosphere 100-fold, de-nitrogenized his discussion of the benefits of guano and ridiculed the experiments of Boussingault. Liebig's patent fertilizers for various crops, released in 1846, turned out to be a fiasco. As the years went by even the students who had thronged to his laboratory in Giessen (Figure 18) became disillusioned by his unwillingness to come to terms with more recent experiments. Several of them discredited his emphasis on soil and ash analysis. Yet his decisive rejection of the humus theory of nutrition and his advocacy of the importance of mineral supplements reoriented theories of plant nutrition, catalysed the application of science to agriculture and created and promulgated an ideal of the research laboratory.

As with Humphry Davy a generation earlier, Justus von Liebig's high standing as a scientist, his incisive reviewing of evidence he had not gathered himself, and his skilful – if sometimes too combative or manipulative – writing ensured that his book had a major impact on farmers as well as scientists. Indeed Sir John Russell says that Liebig's book can fairly claim to be the most important ever published on chemistry in relation to agriculture. One of its unintended lessons is that clarity in science can have two faces, one which stimulates our understanding and one which underestimates the complexity of the problem. Liebig's clarity displayed both.

**Figure 18** Interior view of Liebig's laboratory at Giessen, showing students at work, as illustrated by J.P. Hofmann[100].

## 6.3 Mineral fertilizers and microbial inoculants

At about the same time that Liebig was applying his scientific insight and reputation to a theoretical analysis of agricultural plant nutrition, John Bennet Lawes (1814–1900) of Rothamsted, England was experimenting with simple chemical procedures and field trials for the development of the first effective artificial fertilizer. In spite of Liebig's disdain for field experiments and experimenters, the superphosphate produced by Lawes quickly proved to be far more effective than Liebig's suggested formulation for a phosphatic fertilizer. Clearly there was to be an important role for both theoretical and empirical scientists, and for both laboratory and field experiments, in the application of science to agriculture.

John Bennet Lawes had been a rather casual student at both school and Brasenose College, Oxford but quickly became the energetic and improving squire of his family estate at Rothamsted. Challenged by a neighbouring landowner to explain the erratic response of crops to dressings of crushed bones, he began treating them with sulphuric acid in 1839. So great was the agricultural demand for bones that Liebig accused England of robbing the battlefields of Europe in its ghoulish search for more. Lawes soon extended his acid treatments to mineral phosphates. His field trials with superphosphate showed impressive responses, particularly with turnips, the cornerstone of the Norfolk rotation. His patent, granted in 1842, allowed him to employ a young chemist, J.H. Gilbert, who had trained under Liebig, and to establish a wide-ranging series of long-term experiments.

The public image of the role of Rothamsted at that time was summed up by Philip Pusey: 'The extent of the experimental ground – the expenditure at which it has been kept up – the perseverance with which, year after year, it has been maintained are such as might be expected from a public institution (rather) than a private landowner and render Rothamsted, at present, the principal source of trustworthy scientific information on Agricultural Chemistry.'[179]

With the arrival of Gilbert, the experimental programme shifted away from its earlier focus on superphosphate to a more comprehensive examination of the effects and interactions of all the major nutrients on a variety of crops, pastures and rotations. Their careful experiments made clear the need of most crops – with the puzzling exception of the legumes – for dressings of combined nitrogen, whether nitrate of soda or ammonium sulphate, in the absence of animal manures. Consequently, they vigorously assailed Liebig's assertion that plants obtained their nitrogen from atmospheric ammonia. Liebig responded with equal vigour, but with less constraint by the facts.

In 1857 a young American scientist, Evan Pugh, worked with Lawes and Gilbert on an elaborate experiment to settle the nitrogen controversy. Three cereals and three legumes, as well as buckwheat, were grown in sealed glass structures supplied with air which had been washed through both acid and potash before having carbon dioxide added back. The soil had been sterilized before planting. None of the crops obtained any nitrogen from the air. Their field experiments had led them to expect this with the cereals, but the results with the legumes were puzzling because they had not responded to nitrogenous fertilizers in the field and were known to enrich the soil in nitrogen.

The only problem with this remarkably executed experiment was that they had been too careful in sterilizing the soil. Five years later (in 1862) Pasteur suggested that bacteria could cause nitrification. Consequently when Hellriegel and Wilfarth set out to examine the possibility of atmospheric nitrogen fixation by legumes at Halle in 1886 they not only sterilized the soil as Pugh had done, but added back leachings from a fertile soil to some of the pots. The legumes in the sterile soil soon died whereas those given the leachate soon turned green, grew well and were found to have abundant nodules on their roots, a result quickly confirmed by Gilbert at Rothamsted. The nodules were soon shown by Beijerinck to contain bacteria which could fix atmospheric nitrogen. Hellriegel went on to show that different legumes, clovers and lupins for example, required different strains of the bacterium. It was not long before artificial inoculation with bacterial cultures was suggested, but much more had to be learned about their requirements before artificial inoculation became widespread.

Lawes's Rothamsted Experimental Station, which he set up in trust in 1889, inspired the establishment of agricultural experiment stations in other countries. By the end of the 19th century there were more than a hundred in Germany alone. Following the German example, and with the help of the Morrill Land Grant Act of 1862 and the Hatch Act of 1887, many agricultural experiment stations were also established in the USA.

The legacy of John Bennet Lawes lives on not only in the fertilizer industry which he founded and in the agricultural experiment stations inspired by the example of Rothamsted, but also in the continuing usefulness of Rothamsted's eight 'classic' long-term experiments in assessments of the sustainability of agricultural systems. It is ironic, therefore, that Lawes on one occasion ordered the discontinuance of the Broadbalk wheat experiments on the grounds that nothing more was to be learned from them. Only after an impassioned appeal by Gilbert was the order countermanded.

## 6.4 Vines and verdigris: the chemical control of plant disease

Although diseases of crops often reduced their yields more than fertilizers could increase them, it was not until the late 19th century that the idea of plant diseases being caused by 'microbes' and curable by chemical treatment became at all widely accepted[112]. With insect pests there had been no problem. The treatment of crops with insecticidal preparations had long been practised, with dressings such as the 'pest-averting sulphur' referred to by Homer and the silkworm droppings used by the Chinese in the 4th century AD for the control of rice pests. In the 1860s, Paris Green (copper aceto arsenite) was widely used in the USA to control Colorado Beetle infestations of potato crops.

The problem was that it was generally believed that fungal mildews and blights were the result, not the cause, of plant disease. When the Reverend M.J. Berkeley put forward in 1845 the suggestion that late blight of potatoes was caused by a fungus, his idea was generally ignored, but after Pasteur had proposed his germ theory of disease, the possibility of control by chemicals was more widely entertained. It was appropriate, therefore, that France was the scene of the first successful control of a plant disease by chemical sprays, not surprising that the grape vine was the crop, and predictable that there would be concern about the effects of the spray on the quality of the wine. Downy mildew had reached France from America in 1878 and quickly became the most serious disease of vines. Alexis Millardet (1838–1902) had been commissioned to study it and soon concluded that the best hope of control lay in protecting the vines from the spread of summer spores.

One evening in October 1882 as he was strolling through the

vineyards of St Julien in the Medoc, Millardet noticed that the vine leaves beside the path were not mildewed, although the disease was severe elsewhere. On examining the healthy leaves he found a bluish-white deposit on them, apparently from a chemical treatment. It turned out that it had long been the custom of the Medoc growers to splash the vines alongside paths and highways with verdigris (basic copper acetate), made by covering sheets of copper with fermenting grape husks, to deter theft of the ripe fruit.

'In the field of observation, chance favours only the prepared mind', as Pasteur put it. Millardet recognized that here was a potentially effective treatment for downy mildew. His hopes were raised further by finding that the fungal spores would not germinate in water from his garden well, which was serviced by an old copper pump. However, there was too little mildew on the vines in 1883 to test his treatments, and barely enough in 1884. He compared several compounds of copper and iron, with and without lime, and found the mixture of copper sulphate and lime to be the most effective. The news of his treatment had spread, and others were hot on his heels so he could not wait for yet another season and published his recipe for Bordeaux mixture in May 1885. It was used widely that year, with dramatic success. The reward sought by Millardet for his treatment was neither money nor position but the honour of priority, on which he was insistent because, as he wrote: 'these things are for us *savants* our titles and our most precious trophies.'[112]

Without detracting from Millardet's achievement we should nevertheless note that the fungicidal properties of copper compounds had been proven 75 years earlier by Bénédict Prévost of Montauban. In 1807 he showed that bunt disease of wheat could be controlled by steeping diseased seed in an extremely dilute copper sulphate solution, and that this worked by aborting the germination of the fungal spores. So began the practice of seed disinfection, as effective and as important as Millardet's spraying of leaves. But whatever the complexities of priority, there is no doubt that these crucial experiments, as well as those of Matthew Tillet on bunt in 1750, all stemmed from France and were the foundation of the chemical control of plant diseases. Although discovered by chance and prepared minds, Paris Green, Bordeaux mixture and London Purple, later joined by lead arsenate, remained the mainstays of crop protection for many years.

## 6.5 Crop plant improvement, before and after Mendel

The selection and propagation of superior plants, whether deliberate or incidental, is as old as agriculture itself. The cumulative effects of such selection provided Charles Darwin with crucial insights and evidence for *The Origin of Species*. In its introduction he wrote: 'At the commencement of my observations it seemed to me probable that a careful study of domesticated animals and of cultivated plants would offer the best chance of making out this obscure problem. Nor have I been disappointed ... .'

After the furore following the publication of *The Origin* in 1859 Darwin wrote a masterly summary of *The Variation of Animals and Plants under Domestication* (1868), still the finest account of early plant and animal breeding. But although the changes wrought in domesticated plants and animals by mankind came closest to experimental evidence for evolution, Darwin expressed surprise at 'the little which man has effected, by incessant efforts during thousands of years, in rendering the plants more productive or the grains more nutritious'. Yet he was impressed by the variety of form selected among the brassicas, in which the several cabbages, cauliflower, brussels sprouts, kohlrabi, marrow stemmed kale and curly kale all look very different from their single wild progenitor. He was also impressed by the progressive rise in the sugar content of sugar beet roots. Much emphasis in Darwin's time was placed on the selection of pure lines and varieties, aided by the progeny-testing procedures introduced by the Vilmorin company of France, the earliest seed improvement company.

However, a greater step forward came from early attempts at hybridization. Sex in plants, like sex in Victorian times, was recognized by the practical long before it was acknowledged by the genteel. Early Assyrian farmers knew of the need to pollinate the spathes of date palms to get an abundant harvest, yet it was not until 1694 that Camerarius, in Germany, identified the male and female roles of the stamens and carpels of flowers. Crop plants such as hops, hemp, maize and castor bean provided his evidence yet 70 years elapsed before Kohlreuter produced the first sure hybrid, between two species of tobacco. Recognition that hybridization was easier within species came quickly. From his work with peas, Knight noted in 1823 that the male and female parents made about equal contribution to the first ($F_1$) generation of progeny and that the second ($F_2$) generation was highly variable. Knight also recognized differences between wheat varieties in their reaction to rust, and was the first to cross wheat varieties. Some of the early hybridists also recognized that some characters 'dominate' others, but there was much confusion about how characteristics were inherited, as reflected in Darwin's books.

This was the context within which Gregor Mendel undertook his classic experiments with peas at Brün, the results of which he published in 1866. As is well known, his paper had little impact until it was independently rediscovered in 1900 by Correns, De Vries and von Tschermak. In 1844 plant breeding had been described by John Lindley as 'a game of chance played between man and plants'. It remained so after the rediscovery of Mendel, but plant breeders had much clearer insight into what was happening among their segregating progeny. This was illustrated by Rowland Biffen of Cambridge who was able to show that rust resistance was a recessive character, and that ear morphology, grain colour and baking quality were all Mendelian characters. Although still a game of chance, the odds of success were now much higher thanks to Mendel.

Plant breeding had come a long way from the fumblings of the early pure line selectors and plant hybridists, and its history reveals not only the transforming effects of scientific research on agriculture, but also of agriculture on science, as in the unparalleled example of Charles Darwin.

## 6.6 Daylength and soybeans

On July 10, 1918, as World War I was drawing to a close, H.A. Allard of the United States Department of Agriculture placed three pots of persistently non-flowering Mammoth tobacco plants and a box of Peking soybeans into what he described as a primitive dog house. These plants were to receive only 7 hours of light each day while comparable plants remained in the full summer daylengths of Washington DC.

Allard and his colleague W.W. Garner had decided to do this experiment in the simplest possible way because they had little hope of it succeeding. They had already carried out many experiments varying the light intensity, temperature, mineral nutrition and other factors that might control flowering in these plants, without avail. All their evidence suggested it was a seasonal factor, but they found it difficult to believe that plants could sense and measure the length of the daily light or dark periods. Yet that was what they found. The tobacco and soybeans exposed to long nights promptly flowered, whereas those in the long days and short nights of summer did not. Garner and Allard went on to show that flowering time in many plants is controlled by daylength, and that different varieties of the one crop, e.g. soybean, may respond differently to daylength. Some plants flower fastest in short days, some in long days, while others are indifferent to daylength.

Although they did not realize it, they had in fact been 'scooped' by a young Frenchman, Julien Tournois, who showed in 1913 that precocious flowering of both hops and hemp was induced by exposure to short days or, rather, to long nights. But Tournois was killed in the war the following year. Hops, hemp, soybeans and tobacco are all crop plants, and Tournois, Garner and Allard were all seeking to understand agricultural problems. In so doing they stimulated a great deal of basic research on 'photoperiodism' which has, in turn, led on to more effective crop improvement. I shall illustrate this with soybeans, but the reciprocation between pure and applied research still continues in this field.

The soybean appears to have been domesticated long after the staple cereals. Its wild ancestor still occurs on riverine lowlands in northern China and it was probably domesticated in the 11th century BC[102]. Nevertheless, it became one of the five sacred grains of China. Its use spread to North Korea and to Japan where products such as tofu, tempeh and miso sauce are characteristic elements of the cuisine.

The soybean was first introduced into the USA in 1804, and was grown to a small extent, mostly for hay, in the northern states under seasonal conditions similar to those where it originated. Other varieties of 'the Japan pea', as it was known, came from Commodore Perry's Japan expedition in 1854 but the crop continued to languish until the boll weevil made inroads into the cotton crops of the southern states. In the years preceding the work of Garner and Allard soybeans began to replace cotton in the south, but the varieties adapted to the north flowered too soon, and yields were consequently low. Following Garner and Allard's elucidation of the controlling factor, the many new varieties introduced were classified into eight 'maturity groups', each recommended for a particular latitudinal zone in the USA, Group I the most northerly, Group VIII in the south (Figure 19). The wish to grow

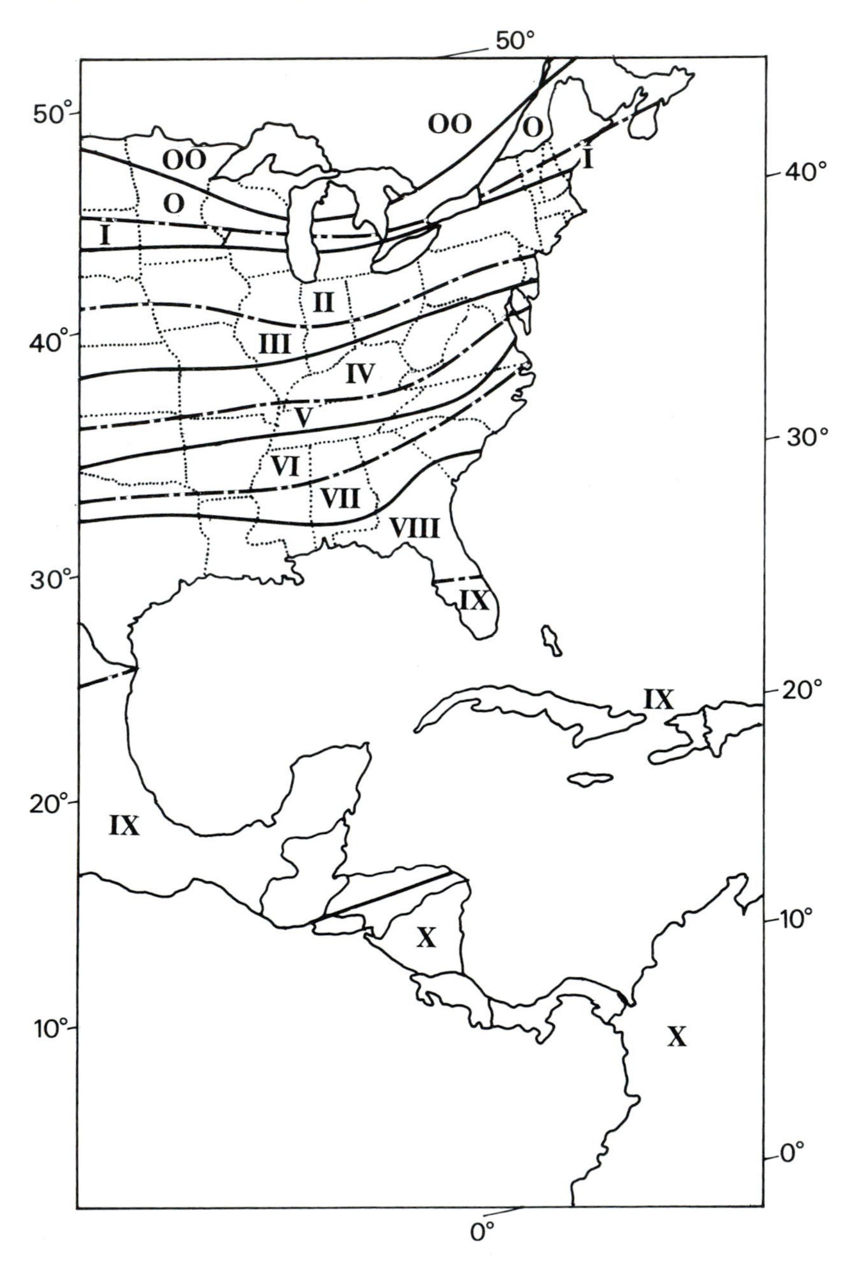


**Figure 19** Zones of best adaptation for cultivars of soybean in maturity groups 00 to X (after Whigham and Minor[220]).

soybeans still closer to the equator has resulted in the addition of two more groups (IX and X) to accommodate suitable varieties.

However, the greatest expansion of soybean crops since the 1930s has been northwards, and has depended on the selection of even earlier-flowering varieties with almost no response to daylength. Groups 0, 00 and 000 now accommodate these. The response to daylength is not only at the first steps to flowering but also through the later stages of the crop, and of course other environmental factors also influence flowering and fruiting. So great has been the expansion of the soybean crop in North America, and so successful the adaptation to the range of seasonal conditions that the USA now produces more than half of the world's crop and 80% of the world trade in it.

In the 1930s Henry Ford, whose enthusiasm for the soybean extended to fuelling his cars and clothing himself with its products, envisaged a vast range of uses for the crop. Some of these are already exploited, and others are on the way, but progress would have been far slower and less sure without insight into the control of flowering by daylength.

## 6.7 The trace nutrient gold rush

In the wake of Liebig's emphasis on the mineral ash content of plants and Lawes' advocacy of nitrogen, phosphorus and potassium fertilizers, there were many experiments on the need by plants for other nutrients. By the 1860s it was known that calcium, magnesium and sulphur were also essential. These were all macro-nutrients, normally present in plants at concentrations greater than 0.1% of dry weight.

However iron, which occurred in plants at a concentration of less than 0.01% of dry weight, was also essential and many investigators sought to be the first to find other essential trace (or micro) nutrients. The favoured technique for this purpose was to grow plants in water culture with the already-known essential nutrients added and test what other elements were needed for healthy growth and completion of the life cycle.

That sounds simple enough, but the requisite high standards of purity of both water and salts were often difficult to achieve with earlier analytical methods. Sufficient traces of the 'trace' elements eluded detection and, in combination with the amounts stored in seeds, often undermined the evidence for essentiality. For many nutrients it was 60 to 90 years after Liebig before widely-accepted proof was obtained. Indeed, the process goes on because it was not until the 1950s that chlorine and sodium were finally shown to be essential, nor until 1983 that nickel was found to be so. The list of essential trace nutrients currently stands at iron, boron, manganese, zinc, copper, molybdenum, chlorine, sodium and nickel. These are available in sufficient quantities in most soils and do not need to be added in fertilizers. But in soils where they are deficient, some crops may show specific deficiency symptoms and respond to dressings of the essential element.

Just as Millardet's prepared mind recognized in 1883 the controlling effect of Bordeaux mixture on downy mildew of the vine, so, in 1911, did Lutman observe that spraying with Bordeaux mixture improved the growth of vines even when they were not diseased. A few years later it was found that spraying with Bordeaux mixture could cure die-back in citrus and the so-called 'reclamation disease' of cereals grown on peat soils. Nevertheless, proof of the essentiality of copper was difficult to obtain because there was sufficient copper for plant growth in the water distilled from the ubiquitous copper stills. Consequently, Winifred Brenchley of Rothamsted found only toxic effects of copper on plant growth when she added it to her solution cultures. Only in 1931 when Anna Sommer of Auburn, Alabama, used specially purified salts and water from a pyrex still was it possible to prove that copper was indeed essential to plant growth.

Much of the early scepticism about the need for copper by plants derived from the fact that even small doses could be toxic. In fact there was only a narrow range between deficiency and toxicity, which is also true for boron. As early as 1910 it had been suggested that boron might be essential. However, the potash fertilizers available in England during the first World War were often toxic to crops, and this was associated with the boron they contained. Winifred Brenchley therefore did some very careful experiments to establish the lower limits of toxicity and the upper limits of stimulation of crop growth by boron. For peas, these limits were respectively, one part in 50,000 and one in 100,000, with a very narrow range between. Brenchley and Kathleen Warington in England and Mazé in France suggested that boron was essential to plants, but once again it was Anna Sommer in Alabama who provided the clinching evidence: the broad beans died in the absence of boron in her experiments.

Early work with manganese suggested that it was not essential for plants, probably because it was a contaminant of the iron salts used in solution culture. However, it was known to stimulate growth in a variety of crops and on a range of soils. In 1897 Bertrand suggested it was probably essential for some catalytic reactions, as also for photosynthesis according to others. That it was essential was established by McHargue of the Kentucky Agricultural Experiment Station in 1922. A few years later, in Australia, it was shown that 'grey speck' disease of oats, symptomatic of manganese deficiency, could be cured in the field by dressings of manganese.

Australia has perhaps the world's most extensive collection of soils deficient in trace nutrients and has provided many confirmations of the curability in the field of the specific symptoms found in solution cultures. For example, molybdenum, by far the heaviest of the mineral nutrients of plants, was shown in 1939 to be essential for solution cultures of barley in which ammonium was the only source of nitrogen. In the following year the symptoms of molybdenum deficiency were found in oats in Australia, as well as a pronounced response by pasture legumes to dressings of as little as 140 grams of the nutrient per hectare.

Of all the trace nutrient deficiencies, however, zinc – also shown to be essential by Anna Sommer – is the one which currently requires the most extensive amendment by fertilizers. In the USA this amounts to about 40,000 tonnes each year, three times as much as manganese, 40 times as much as copper and 400 times as much as molybdenum, but only one 250th as much as nitrogen.

The trace nutrients clearly live up to their name. But although the scale of their application is small in comparison with the major nutrients, they can have a substantial effect not only on crop and pasture yields but also on extension of the area under cultivation and on the nutritional quality of the harvested crops. Indeed all the nutrients essential for the growth and development of plants, with the exception of boron, are also needed by farm animals, together with a further nine which plants accumulate without needing. These are the Davids which can render the Goliaths of nitrogen, phosphorus and potassium helpless in both plant and animal husbandry.

## 6.8 Biological control of pests and weeds

In 1800 Charles Darwin's grandfather Erasmus, in his book *Phytologia*, suggested that predatory fly larvae could be used to control aphids in hothouses. Long before that mynah birds had been introduced to control locusts, and the ancient Chinese are known to have fostered ants to control caterpillars and boring beetles on citrus, even providing bamboo runways for the ants to move from tree to tree.

The term 'biological control' was introduced in 1919 for the control or regulation of pest populations by natural enemies i.e. predators, parasites or pathogens. Some degree of biological control is operating naturally in most crops. Indeed this internecine warfare is extremely important to crop health and productivity, and it is ironic that its impact on pests that have become resistant to DDT is best exposed by spraying with that insecticide to kill off their natural enemies, as in rice infested with brown plant hoppers.

Here, however, we are concerned with the consequences of the introduction of crops to new regions. In quite a few instances they left their pests behind, at least initially, but in many cases the pests accompanied or caught up with the crop but left their own natural enemies behind. In such circumstances the pests may cause great damage to the crop, threatening its usefulness in its new environment and inspiring a search for its natural enemies.

The first spectacularly successful example of this kind of biological control was in California in 1889, the pest being the cottony-cushion scale which was devastating the citrus industry of the state. The scale had first appeared in California in 1868 and had probably come from Australia. Fumigation and the insecticides of the time were ineffective. C.V. Riley, the chief entomologist of the US Department of Agriculture (USDA), proposed sending one of his staff to Australia to search for natural enemies of the scale, but Congress refused to allow USDA funds to be spent on foreign travel, nor would the State government provide them. Finally, through a lateral arabesque, Albert Koebele was sent to Australia in September 1888 ostensibly to represent the US State Department at the Melbourne International Exposition.

By October Koebele had discovered an effective predator of the scale, the ladybird beetle *Vedalia*. This was shipped to California, and quickly released in two commercial orchards. By April 1889 the ladybird population was exploding and the scale was being eliminated. Hundreds of citrus growers came to collect the ladybird from the two trial orchards. By 1890 the cottony-cushion scale was no longer of any consequence, until 70 years later, that is (cf. p. 160). The benefits from a total expenditure of about $5,000 amounted to millions of dollars annually[50], but more than that, there developed a uniquely American relationship between agriculture, government and entomology.

Finding such an effective control agent so quickly has proved to be rare. In the case of the sugar cane leaf hopper which first appeared in Hawaii in 1900 and was seriously damaging the crop two years later, the search for natural enemies began, again in Australia, in 1904 but was not consummated until 1920. Five parasites of the leaf hopper were found in Australia, Fiji, China and Taiwan, and tested in Hawaii, but had little effect. Finally, Frederick Muir, puzzled by some fungal infections of the eggs of the leaf hopper in Queensland, discovered that these occurred in eggs that had been sucked by a predator from an unusual group of insects. As Muir put it: 'the least obvious of the death factors was the keystone of the complex and one that was overlooked by observers in different parts of the world'. Released in Hawaii in 1920, it had reduced the leaf hopper to insignificance within three years.

An early example of effective biological control of weeds was provided by the prickly pear cactus (*Opuntia*). Introduced early into Australia, it escaped from cultivation in the 1800s and by 1925 it infested 25 million hectares of agricultural land so heavily that they were impenetrable and useless. Searches for natural enemies of the cactus in the USA, Mexico and Argentina resulted in over 50 insects being sent to quarantine nurseries in Australia. One of these, sent from Argentina early in 1925, was *Cactoblastis cactorum*. Alan Dodd, who collected it, saw no reason to believe it would be any more successful than the others, yet only four years after its release in 1926 the prickly pear was no longer a serious weed.

Other examples of successful biological control both of insect pests and of serious weeds could be given. The successes in pest control have been mainly with perennial plants such as tree crops, stone fruits, olives and particularly citrus, as well as introduced forest trees. Pests of sugar cane, in the ground for two years, have also proved suitable for biological control and there are possibilities with rice particularly in areas where two or three crops are grown each year. Biological control of stem borer in rice has reduced yield losses from that pest in Hawaii and the Philippines, for example.

But the life cycles and management of the major food grains tend to limit the effectiveness of biological control. Their short life cycles and crop rotation practices create changing environments which limit the conservation and increase of natural enemy populations. The relatively low value per hectare of the food crops tends to make repeated releases of parasites and predators uneconomical. Nevertheless, there have been some successes, such as the reduction of injury by the European wheat stem sawfly by the acclimatization of an ichneumonid in Canada.

Introduced enemies now control more than 70 species of weedy plants, mainly in rangelands and perennial crops[6]. Their effectiveness on weeds of the staple food crops is subject to the same limitations as apply to the crops themselves. Nevertheless, biological control was an exciting and valuable agricultural innovation, one which deploys our understanding of the subtleties of the web of life. There are still many opportunities for its application, and for its full flowering as the key to integrated pest management (p. 160).

## 6.9 Botanic gardens and plant introductions

Ever since the first farmers took their crops with them through Europe, India and Africa, explorers, navigators and colonists have both introduced and collected useful plants along their way. Columbus, for example, brought back maize and sweet potato, which he presented to Queen Isabella, and returned to the Americas the following year with sugar cane and citrus, both previously brought to Spain by the Arabs. Crop plants have long been a global resource. For most of them, their centres of production are now far removed from their centres of origin.

Botanic gardens played an important role in the post-Columbus redistribution of plants. The first to be established was at Pisa in 1543, closely followed by Padua and subsequently by many other European cities. The initial impetus for these gardens was the need for the burgeoning medical schools to have access to collections of medicinal herbs. Such a collection, of Chinese medicinals, had been established as early as 2800 BC by the Emperor Sheng Nong.

From the 16th century on, however, the competing colonial powers of Europe began to use their botanic gardens for the collection, study, propagation and dissemination of agricultural, horticultural and medicinal plants of value to themselves or to their colonies. The first botanic garden in the tropics was established by the French at Pamplemousses on Mauritius in 1735. Cassava and vegetables were planted, partly to provide food for the French community, and then fruit trees including grapefruit (hence the name of the gardens). By 1767 plantation spice crops (such as nutmeg, cinnamon and pepper) and various tropical fruits and dyes were under test.

The Royal Botanic Gardens at Kew in London came late on the scene, being established only in 1759, but the scope of their ultimate impact owed much to the early travels, broad vision and great influence of Sir Joseph Banks. He foresaw Kew as playing an important role not only in the expansion of the British empire but also in studies of the flora of the world and in promoting human welfare. He personally financed many plant collecting expeditions, and recognized the potential value of Kew's network of botanic gardens in the tropics for the redistribution and naturalization of crop plants around the globe.

The St Vincent Botanic Garden (established 1766) was, for example, the way-station for the introduction of the breadfruit to the New World in 1793 after Captain Bligh's first attempt four years earlier had failed with the mutiny on the *Bounty*. Nutmeg from Asia and sugar cane from the Pacific also entered the Americas via the St Vincent gardens.

Kew's satellite gardens in Peradeniya (Sri Lanka) and Singapore played an equally important role in introductions in the reverse direction of New World crops such as rubber, coffee and cinchona into Asia. Some writers, such as Lucile Brockway in *Science and Colonial Expansion*, have accused Kew of using its network of botanic gardens in an underhand way on behalf of the British empire. Her accusation has been discredited but her conclusion 'that Kew Gardens and its affiliates had an important role in empire-building by virtue of scientific research and the development of economically useful plants for production on the plantations of the colonial possessions' stands. While Britain gained some advantages, so also did the developing countries. The introduction of rubber continues to strengthen the economic independence of Malaya and Indonesia, and many countries have benefitted from the introduction of cinchona to India, just as South Americans have from the introduction of wheat, rice, bananas, sugar cane and many other crops.

Indeed, many countries now have national agencies for plant introduction, following the lead of the United States. The US Department of Agriculture, established in 1862, set up its Office of Foreign Seed and Plant Introduction in 1898. The political defeat of the cattle ranchers of the north-west plains had created a need for cold-tolerant crops which was met partly by the introduction of many hard red winter wheat

**Figure 20** N.I. Vavilov (right) during his exploration of Abyssinia in 1927 (Lenin Academy of Agricultural Science[116]).

varieties. In the south-east more drought-tolerant crops were the requirement, a need met partly by the introduction of *Acala* cotton from Mexico. In the longer term, however, the greatest success of the Office of Foreign Seed was the establishment of soybeans as a major American crop as a result of the introduction of a great variety of Asian accessions.

But of all the adventurous and intrepid plant collectors, Nikolai Ivanovich Vavilov (1887–1943) explored the most widely (in all continents except Australia), generalized the most boldly and, through his All-Union Institute of Plant Industry, built the most comprehensive system for evaluating, studying and using the 200 K introductions from over 70 expeditions. Vavilov personally undertook many of the most daunting and dangerous of these, such as Persia in 1916, Afghanistan in 1924, Abyssinia in 1926/7 (Figure 20), China and Korea in 1929 and Central and South America in 1930/1 and again in 1932/3, himself collecting more than 60 K samples[160].

Vavilov may well have been in Abyssinia when the world population first reached two billion. His dream, at that time, was of a world collection of useful plants with 'the possibility that in the near future man will be able, by means of crossing, to synthesize forms such as are absolutely unknown in nature', a dream now closer to realization.

CHAPTER 7

# The third billion (1927–1960)

## 7.1 Introduction: inputs and interactions

Adding the second billion to the population of the world had taken a century whereas the third billion took only a third as long. Nevertheless, those 33 years encompassed several major innovations and laid the foundations for a profound change in world agriculture.

In spite of this, however, by far the most important source of the increase in food supply for the 50% increase in population was still the age-old one of increase in the area cleared for cultivation. Even in the 1950s there was a substantial expansion of the arable area in the Soviet Union, with the ploughing up of 36 M hectares of steppe. Between 1927 and 1960 the global arable area was extended from about one billion hectares to 1.4 billion (Figure 17, p. 91). To this increase of 40% we should add an amount for the land no longer used to grow food for draught animals, which may have been about 0.13 billion hectares, making the effective increase in arable for the period about 53%. The additional arable land may have been of poorer quality than the previous crop land, but even so it is clear that extension of the arable remained the dominant contributor to increases in world food supply until 1960. Since then the gains in arable area have been largely balanced by the losses, and the food requirements of the growing population have been met by rising world average yields. Reaching a population of three billion in 1960 was, therefore, a major watershed in world agriculture.

Production statistics published by the United Nations Food and Agriculture Organization (FAO) and its precursor indicate that from the mid-1930s to 1960 the world average wheat yields rose by only about

12%, and rice yields by only 9%, as the population increased from two to three billion, although average maize yields rose 55%, reflecting their rapid increase (of 120%) in the USA. The 1927–1960 period saw the establishment of the FAO and its annual publication of the agricultural statistics on which our global housekeeping can be more securely based. It also witnessed the Universal Declaration of Human Rights by the United Nations in 1948, including that to an adequate food supply.

Besides reaching the end, at least for the time being, of the increase in arable area, this period is characterized by four other major transformations of agriculture. The first is the increasing dependence on off-farm inputs, still very small in 1927. Since then liquid fuel and electricity have largely replaced men and horses as the source of energy for farm operations as a result of mechanization. Nitrogenous and other fertilizers increasingly replaced farmyard manure. The revolutionary insecticides, fungicides and herbicides quickly came to be relied on for pest and weed control. Even for the seed they sowed, farmers began to look beyond the farm gate, as they had to do for hybrid maize.

Secondly, along with this change went an increasing 'industrial appropriation of agricultural inputs', as Goodman, Sorj and Wilkinson[72] put it. They saw the uniqueness of earlier agriculture as having been its confrontation of capitalism with a natural production process to which there was no industrial alternative. However, the innovations discussed in this chapter, and others like them, have led to a flourishing of agribusiness. Mechanization, initially the outcome of ingenious inventions by many farmers, has been taken over by industry, as has the supply of fertilizers and other agrichemicals. The rapid adoption of hybrid maize hastened the industrial appropriation of seed supplies. At the same time, this penetration of agriculture also promoted industrial research for agriculture, such as that which led to DDT and many other agrichemicals.

Thirdly, the development of cheap nitrogenous fertilizers for crops made it possible for farmers to abandon the mixed crop/animal husbandry systems which represented the earlier ideal of agriculture – such as the Norfolk four-course rotation – for more specialized crop production systems.

Fourthly, mechanization and the industrialization of agricultural inputs created a great divide in world agriculture between the more developed countries where rising crop yields meant that extension of the arable could be limited, and the less developed ones still in their demographic transition and with rapidly increasing populations which needed continued land clearing to meet their requirements for food[77].

Although the subjects discussed in this chapter were chosen because of their ultimate impact on agriculture, they also illustrate the great variety of routes to agricultural innovation. Mechanization required ingenuity and persistence more than research, by many farmers and later by small machinery companies. Industrial research led to cheap nitrogenous fertilizers and to Paul Müller's discovery of the insecticidal properties of DDT. Basic research revealed the phenomenon of hybrid vigour in maize, which was then developed by the commercial sector. Basic research also led to 2,4-D and thence to the herbicide industry. Plant tissue culture is a particularly interesting example of the unpredictable impact of basic research in that although even its early proponents saw little scope for application, its uses continue to proliferate, one of them being as the Trojan horse for the genetic engineering of crop plants.

Clearly, many kinds of research contribute to the improvement of agriculture. Advances in one sector often prove to be synergistic with those in others, as were mechanization, nitrogenous fertilizers, herbicides and the breeding of hybrid maize. Moreover, innovations like DDT, 2,4-D and cheap fertilizers which are seen initially as major advances can later become problematic as unforeseen side-effects are exposed. But that is true also of the basic agricultural act, the clearing of land for crops which has, since time immemorial, often led to soil erosion. The answer is not to abandon agriculture, or fertilizers, or agrichemicals altogether, but to search for solutions to the problems. On occasion those solutions may require the recognition of public responsibilities and legislative action, as the example of the US Soil Conservation Service shows.

Finally, as in our earlier steps up the population ladder, famines were still not banished; witness Bengal in 1943 and China in 1959–61 when 30 million people are believed to have died of starvation. Pessimists such as William Vogt in his *Road to Survival* (1948) were convinced that global food production and population were already close to their limits. One group of experienced and knowledgeable agricultural scientists asked in 1951: 'How shall we work the miracle of feeding the 4000 millions', and emerged uncertain of their answer[114]. Yet there are now many more than four billion of us and, on average, we are better fed than in 1951 thanks to the continuing impact of the innovations discussed in this and other chapters.

## 7.2 Mechanization replaces men and horses

At the annual Farmers and Homemakers Week meeting in Minnesota in January 1927 an Iowa farmer, Roy E. Murphy, described his 200 acre 'horseless farm' to a highly sceptical audience. At that time the rural workforce was as large as it had been at the end of the 19th century and there were still 22 M horses but only 600 K tractors on American farms (Figure 21). Yet almost 70 years before that President Abraham Lincoln had said: 'The successful application of steam power to farm work is a desideratum ... it must ... plow better than can be done by animal power. It must do all the work as well, and cheaper, or more rapidly, so as to get through more perfectly in season.'[54]

There in a nutshell was the agenda for mechanization, one to which many American farmers and farm businesses contributed. The tractor was the key to it all. By 1960 the man-hours of farm work had been reduced by half and there were few horses but almost five million tractors on American farms.

Abe Lincoln's 'desideratum' had to await the development of the

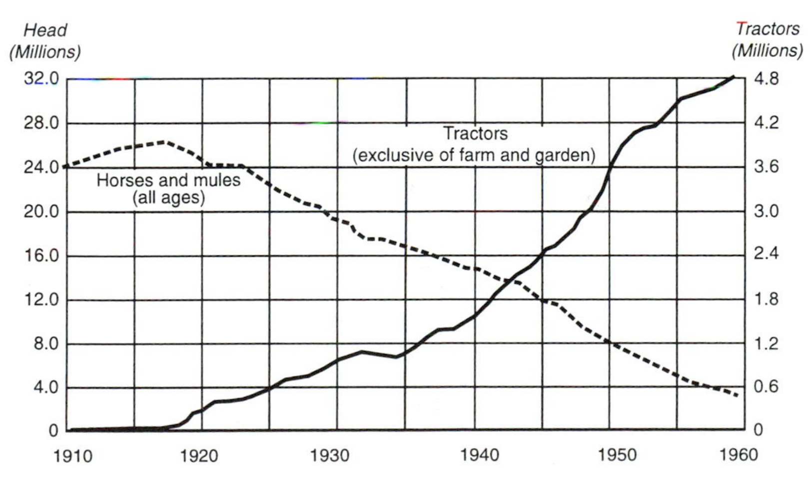


**Figure 21** The transition from horses and mules to tractors as a source of farmpower in the USA (Dieffenbach and Gray[54]).

internal combustion engine for its realization. Only then could the many 19th century inventions of reapers, binders, threshers and traction engines be combined in an effective and affordable way. But much inventive ingenuity by farm folk had laid the foundation.

Sowing in rows by drill rather than by 'the immemorial arc' of the farmer's arm was one of the earliest attempts at mechanization. Jethro Tull's drill was introduced around 1700 but remained suspect and not widely used in England until the 1830s. More success attended the winnowing machine introduced from the Netherlands into Scotland in 1710[170]. This led on to Andrew Meikle's invention and patenting in 1788 of the first 'portable threshing machine', which was quickly adopted in Scotland and the north of England, the traditional flail being laid aside.

The scene then shifted predominantly to the USA where, in 1831, Cyrus McCormick invented the first widely-used reaper, in the wake of several earlier attempts in France and England. From then on patenting became highly competitive, and many farm machinery companies were formed. In 1836 Moore and Hascall began their prolonged attempt to develop a horse-drawn combination for harvesting and threshing, first in Michigan and, from 1854 on, in California where the drier climate was more suitable. After a successful demonstration of its speed and economy there, however, one combine was burned by farm workers fearful of its effects on rural employment, a concern that faced many innovators of mechanization. Several standing steam threshing machines were introduced by 1850, and a mobile model by Henry Ames in 1854. Although these were greatly improved up to the 1880s, they still required a small army of labourers to accompany them.

Self-raking reapers were introduced in 1854, each requiring one less man for harvest, and became predominant within 10 years. Steam ploughs were introduced in 1868 and disc harrows, wire binders and then twine binders in the 1870s. Mobile steam-powered combines began to be used in California, as vividly described in 1878 by Professor Hilgard: a 'wondrous and fearful combination of header, thresher, and sacking-wagon moving in a procession side by side through the doomed grain'. He went on to suggest that 'we shall doubtless see the flouring mill itself form a part of the pageant'. Such was the fascination with mechanizing combinations of functions at the time. However, the use of combines did not spread to the eastern USA until the 1930s. The key was the development of the internal combustion engine and its integration with a smaller, more mobile tractor body.

The gasoline tractor was introduced in 1892 by J. Froelich, and the first model with tracks by Benjamin Holt in 1907. In the following year Henry Ford, a farm boy from Michigan, first tried to combine traction wheels with his model B chassis. By 1917, when he introduced the Fordson tractor, 200 companies had already produced 50 K tractors in the USA. But his were smaller and cheaper and could replace horses for many lighter tasks just at the time when large numbers of horses were being sent to Europe following the USA's entry into the war. By 1925 Fordson tractors had captured three-quarters of the American market, the gasoline-powered tractor was launched into its key role in farm mechanization, and the horse made its exit, lingering on only in the power ratings of the tractors and in the space between rows of maize, the width of a horse apart.

The impact of these beginnings on agriculture was particularly marked in America and Europe although it continues to expand both there and in the rest of the world. As the power per rural worker increased, not only from 'tractorization' but also from rural electrification, much labour was freed for industrial development while output per farm worker and per hectare rose sharply. Thanks to the increased speed of operations, their timeliness was improved, as Lincoln had foreseen. There was also a rapid increase in the area of arable land available for food production, by up to one third, as horses – each of which required about one hectare of crop for feed – were replaced by tractors. On the other hand, agriculture became highly dependent on external energy and sensitive to changes in its supply and price.

Taken overall, the mechanization of farm operations was among the most influential of all advances in agriculture. It was achieved by the ingenuity and persistence of a great many farmers, blacksmiths and mechanics without much benefit of science until more recent times.

## 7.3 Cheaper nitrogenous fertilizers

Just one hundred years after the publication of Malthus' *Essay*, the English physicist Sir William Crookes suggested, in his presidential lecture to the British Association in 1898, 'that the fixation of atmospheric nitrogen could be the chemist's contribution to the maintenance of civilizations based on wheat'. That advance was realized sooner than he expected, to the benefit of many crops besides wheat.

Nitrogen and oxygen were combined over an electric arc at high temperatures to yield oxides of nitrogen in Norway in 1905. The following year calcium cyanamide was formed from nitrogen and calcium carbide at high temperatures, a process used briefly for fertilizer production in England and Japan. Then in 1908 the combination of nitrogen and hydrogen to form ammonia at high temperature and pressure in the presence of a catalyst – the Haber–Bosch process – was discovered, and the first factory to use it was built in Germany in 1913. This process is still the basis of modern production of nitrogenous fertilizers, and subsequent industrial refinements have brought manufacture close to its theoretical efficiency and have greatly reduced its cost. Nitrogen is derived from the air and hydrogen from a variety of sources, mainly natural gas in the USA, coal and coke in Japan.

Ammonia is now the principal source of nitrogen in fertilizers based on nitrates of calcium, potassium, sodium and ammonium, as well as in urea-based and nitro-phosphatic fertilizers. As their availability has increased and their relative cost has diminished, nitrogenous fertilizers have come to dominate the scene. Until the 1960s world production of fertilizer nitrogen (N) was comparable in amount with that of potassium (as $K_2O$) and phosphorus (as $P_2O_5$), the somewhat arbitrary bases for all FAO fertilizer statistics. Now, however, almost three times as much fertilizer N as of $P_2O_5$ or $K_2O$ is consumed. The total world use of more than 80 M tonnes of N each year is now only 20–30% less than the total amount of terrestrial biological nitrogen fixation. Its scale is vast, and its impact extends far beyond agriculture, to the whole biosphere. Vaclav Smil[194] estimates the annual flux of N in global croplands to be 175 M tonnes (t), of which fertilizers provide almost half compared with 30 M t from mineralization, 30 M t from biological fixation and only 15 M t from animal wastes.

In the USA, unlike the rest of the world, more than half of the nitrogen in fertilizers is used as ammonia, either dissolved in water or anhydrous, as the compressed and liquefied gas. Although ammonia solutions were experimented with in the 1850s, it was not until 80 years later that they were first used for flood irrigation of citrus plantations in California by D.D. Waynick. Anhydrous ammonia was first applied directly to soil in 1930, and by the 1940s W.B. Andrews of the Mississippi Agricultural Experiment Station had developed the system of application now widely used and based on a tractor-mounted cylinder, flow meter and knife-like applicator which inserts the ammonia at a depth of about 15 cm in the soil. Since then the use of ammonia as a fertilizer has expanded spectacularly in North America and will presumably spread more widely round the world in future, given that it is highly efficient and costs only about half as much as solid nitrogenous fertilizers.

The great surge in agricultural use of nitrogenous fertilizers has had a huge impact on global food production, as is evident in the parallel rises in N fertilizer use and world average yields of wheat and rice since 1960 (Figure 17, p. 91). It was driven by technological advances in the fertilizer industry and their economic consequences more than by advances in agricultural research. Moreover, to a considerable extent these advances caught agriculturists by surprise, as Neal Jensen[108] has admitted in the case of wheat breeders: 'they did not conceive of a real world in which (tall-strawed varieties) would no longer be suitable – a world with inexpensive and abundant nitrogen supplies ...' As so often, the innovation was slow and uncertain in its initial impact, and there was little basis in past experience for agriculturists to foresee the prerequisites for securing the full advantages of cheaper nitrogenous fertilizers.

## 7.4 Hybrid maize: society wedding or shotgun marriage?

In Shakespeare's play *The Winter's Tale*, when Perdita spurns the hybrid carnations 'which some call nature's bastards', Polixenes replies 'This is an art which does mend nature – change it rather – but the art itself is nature'.

In early maize crops that art was indeed natural, given both the open-pollinated breeding system of maize and the religious and agronomic customs of its early cultivators, which would have conferred a high level of hybridity. Deliberate hybridization was first reported for maize in 1877 by Charles Darwin and by W.J. Beal, who both found hybrids to be more vigorous than inbreds. Many yield comparisons followed and in his review of them in 1922 F.D. Richey found that in about half of the crosses the yield of the hybrid exceeded that of the higher-yielding parent, although not usually by a wide margin.

George Shull of the Carnegie Institution at Cold Spring Harbour, New York shaped the new approach. Shull's initial interest was not in plant breeding as such but in the number of kernel rows per ear as a quantitative character. From these basic studies he concluded: '1) that in an ordinary field of corn, the individuals are very complex hybrids; 2) that the deterioration which takes place as a result of self-fertilization is due to the gradual reduction of the strain to a homozygous condition; 3) that the object of the corn breeder should not be to find the best pure line but to find and maintain the best hybrid combination.' Although this flew in the face of contemporary plant breeding procedures for other crops, Shull had diagnosed the way ahead. He later commented that his proposed method, though very simple, would be a 'somewhat costly process.'[196]

One reason for this was that the inbreds produced so little seed, but Shull's proposed breeding programme became practicable in 1918 when D.F. Jones suggested the use of double cross hybrids from high-yielding single crosses for seed production. This got around the conclusion of his Harvard professor, E.M. East, that the pure line method was not commercially feasible, although it was not known at the time that only a rare combination of single crosses blend effectively into a highly productive double cross.

However, the use of hybrid maize remained quite limited until the 1930s (Figure 22). It was not promoted strongly by the state agricultural experiment stations, where many plant breeders were more impressed by the injurious effects of inbreeding than by the positive gains from hybridization. The US Department of Agriculture was slow to promote hybrids until F.D. Richey joined the staff under Secretary Henry C. Wallace. The latter's son Henry A. Wallace, later to be also Secretary of Agriculture, was an enthusiastic promoter of hybrid corn, selling the first commercial hybrid, 'Copper Cross', in 1924 and founding the Pioneer Hy-Bred Corn Co. of Iowa in 1926.

It was the conjunction of three trends that eventually led to the rapid adoption of hybrid maize from less than 1% in 1933 to about half of the total crop in the USA 10 years later. The first was the establishment of many small maize breeding companies producing commercial hybrids adapted to local conditions from proprietary inbred lines. Secondly, the uniformity of these hybrids, especially in their growth and the position of their cobs, encouraged the development of mechanized harvesting, thereby reducing labour costs. And thirdly, fertilizer application to maize crops at last began to increase following the exhaustion of the natural fertility of the prairie soils which had led to declining yields

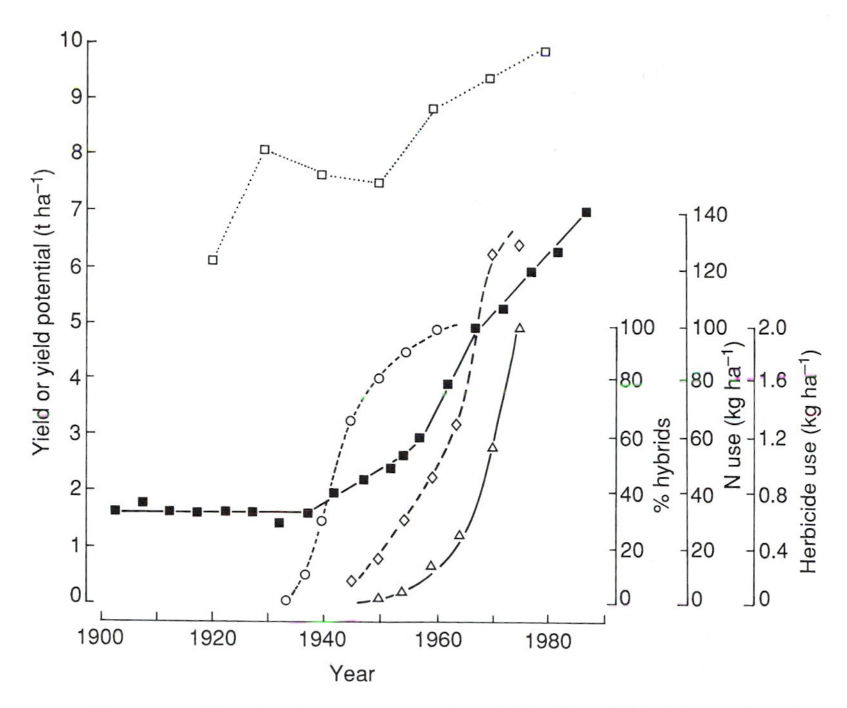


**Figure 22** The rise in average maize yield of the USA (■) as related to the percentage of area sown to hybrids (○), the use of N fertilizers (◇) and herbicides (△), and the improvement in yield potential by breeding (□)[58].

until the mid-1930s. As so often in the history of agriculture, it was the synergistic effects of several interacting innovations that led to rapid change.

The proprietary inbreds secured the profits which led to further investment in the breeding of superior hybrids and this in turn encouraged the wider adoption of hybrids to near-completion by 1960, greater use of fertilizers, herbicides, irrigation and other inputs, and a steady rise in yield (Figure 22). The average yield of maize in the USA is now four times higher than it was in the 1930s, and the yield potential of current Pioneer hybrids is about twice as high as that of their first hybrids, mainly due to their better adaptation to the much closer planting practised since mechanization.

There could have been other roads to successful improvement of maize. As Don Duvick[55] says: 'the breeding methods for population improvement that are now available might be as effective as the inbred-hybrid method for making certain kinds of genetic gains'. But that is today, not yesterday, and the history of hybrid maize in mid-west America illustrates the powerful impact of strong investment in plant breeding on other components of agricultural innovation.

## 7.5 A heritage of erosion, a hope of conservation

On May 12, 1934 the first dust cloud large enough to cross the USA intact from the parched fields of Kansas, Texas and Oklahoma passed over New York and Washington and swept out to sea. The darkened skies shocked people on the eastern seaboard with the realization that something had gone wrong with the land to the west. In fact, about one third of the arable land of the USA was under threat of erosion at the time.

Soil erosion is as old as agriculture itself, and its history is an integral part of the history of agriculture. We saw earlier that the Greeks recognized its causes and its effects, yet did little about it. In the USA William Byrd of Virginia had recorded severe loss of soil from his tobacco fields as early as 1685. In 1907 when Gifford Pinchot initiated the movement for the conservation of natural resources, soils were among them. However, it was a soil survey in South Carolina in 1911 that provided the first compelling evidence of the extent of erosion on agricultural land[12].

In 1933 a Soil Erosion Service was established within the US Department of the Interior, but the great American crusader Hugh Hammond Bennett wanted a body that would survive the Depression and commit the federal government to a policy of long-term soil conservation. In early April 1935, when he knew another dust storm was on the way, Hammond appeared before the Senate Public Lands Committee just as the dust from the plains arrived. The Senators suspended the hearing for a moment as the sky darkened, and moved to the windows to watch[92]. Soil conservation was clearly a public responsibility. The preamble to the Soil Conservation Act of April 27, 1935 which ensued reads: 'It is hereby recognized that the wastage of soil and moisture resources on farm, grazing, and forest lands of the Nation, resulting from soil erosion, is a menace to the national welfare and that it is hereby declared to be the policy of Congress to provide permanently for the control and prevention of soil erosion and thereby to preserve natural resources, control floods, prevent impairment of reservoirs ... protect public health, public lands and relieve unemployment.'

The establishment of the Soil Conservation Service under Bennett allowed a comprehensive approach to the problems of soil erosion. Research into methods of control, across a wide range of environmental conditions, could be combined with cooperative activities with farmers, broadly based demonstration and educational programmes, and legislative initiatives. The states were encouraged to pass Soil Conservation District laws. Following enactment of these the Soil Conservation Service could provide expert personnel to each district, to work with one another and with farmers on demonstration programmes, which resulted in effective technology transfer and less adversarial relations.

Many interlocking techniques were developed and promoted: contour ploughing, an old practice insisted on by Columella and other Roman writers but weakened by the advent of the tractor; strip cropping; stubble mulching; wind breaks; subsurface tillage; and greater use of 'permanent' pastures. Consequently, when the drought returned in the 1950s, although there was some wind erosion, there were no black blizzards.

Better rangeland management techniques were developed and the condition of the Dust Bowl rangelands improved. On cropped land, the improvements in equipment for cultivation and in techniques for crop residue management have had a major impact on soil conservation, particularly with the development of reduced tillage, ridge-till, strip-till, mulch-till and zero-till techniques. Although farm machinery improvement has been crucial for this, so has the introduction of herbicides. By 1988 almost a quarter of the southern plains and a third of the northern plains were planted by conservation tillage. Wind erosion has been greatly reduced, but not entirely banished from the Great Plains, as the dust storm of 1989 indicated.

Nevertheless, Hugh Hammond Bennett's dream of an investigative, cooperative and educational Soil Conservation Service with long-term federal government support has been vindicated. The area of arable land in the USA is 12% greater now than it was in 1930. Faulkner's indictment of the *Plowman's Folly* is occasionally merited when commodity price support programmes or high prices on the international markets tempt farmers to crop areas beyond the long term capacity of their land but, so far at least, *The Grapes of Wrath* have been avoided.

## 7.6 Hormones as herbicides: the discovery of 2,4-D

The discovery of the herbicidal action of 2,4-dichlorophenoxyacetic acid (2,4-D) is a telling example of the agricultural value of research driven initially by curiosity alone. Within a few years of their discovery, the hormone herbicides had become an important input, with wide-ranging agricultural and social impacts.

Along with pests and diseases, weeds had always been a problem in agriculture but may, to a degree, have been tolerated as inevitable. The old writers say little about them and, apparently, neither the Greek nor the Latin languages had a specific word for weed despite their rich vocabularies[179]. Hand-hoeing or repeated cultivation were the predominant means of weed control until the 1880s, when copper sulphate was found in France to be toxic to some broad-leafed weeds as it was to downy mildew of vines. In the 1920s, sodium chlorate and perchlorate were added to this toxic, inflammable and corrosive armoury of inorganic herbicides. In the 1930s derivatives of benzene, especially dinitro-orthocresol, were also found to be partially selective herbicides, killing broad-leafed weeds but not the cereal crops when used at doses of only 6–10 kilograms per hectare. By then, however, the hormone herbicides, more selective and ten times more active, were on the verge of discovery.

Among the last experiments done by Charles Darwin, with the help of his son Frank, were some ingenious ones with hooded or painted seedlings of cereals and grasses from which he concluded that 'some influence is transmitted from the upper to the lower part, causing the latter to bend'. This highly original deduction was strongly opposed at first, but at 3 a.m. on April 17, 1926 Frits W. Went obtained the first evidence that Darwin had come to the right conclusion in 1880. At the time, Went was working towards a doctorate in his father's laboratory at Utrecht and was so excited that he ran home, woke his father and asked him to 'Come and see. I've got the growth substance.' His father told him to repeat the experiment, and went back to sleep. The next day he repeated the experiment to his father's satisfaction, and went on to isolate the *wuchsstoff* and demonstrate its hormonal nature and polar transport.

Then followed a confusing and distressing interlude during which the hormone was at first wrongly identified by a team of chemists at

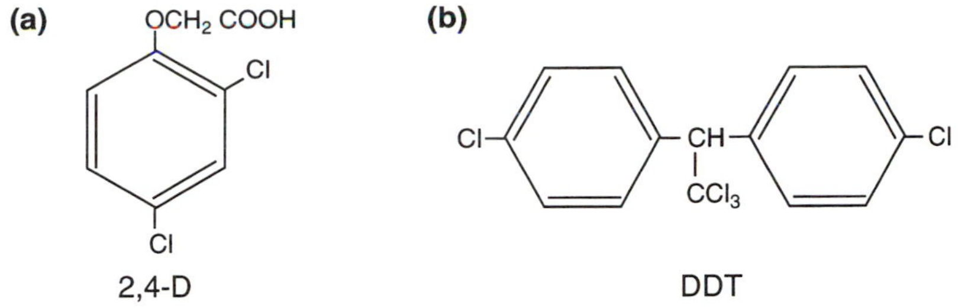

(a) 2,4-D

(b) DDT

**Figure 23** Molecular structures of 2,4-D and DDT.

Utrecht who eventually, after having thrown an American group off the indole trail, identified it correctly in 1934 as indolyl-3-acetic acid (IAA)[224]. An avalanche of plant physiological experiments in the 1930s left few aspects of plant life apparently untouched by this hormone.

The idea of using related synthetic hormones as weed killers evolved independently in several English and American groups as World War II gathered momentum, preventing the full and sequential publication of their results because of security restrictions. The selective herbicidal action of IAA and related compounds on charlock in a crop of oats was first shown at Jealott's Hill in England in 1940[206]. The following year a simple procedure for the synthesis of substituted phenoxyacetic acids such as 2,4-D (Figure 23) was published and several of these compounds were shown to have a wide spectrum of physiological effects on plants. In the latter part of 1941 E.J. Kraus, at Chicago, suggested to some colleagues that these growth regulators might work as herbicides[153]. Quite independently, three research groups in England found that 2-methyl-4 chloro-phenoxyacetic acid (MCPA) was also an effective and selective herbicide for cereal crops. These three groups were finally brought into contact with one another under an umbrella of secrecy in 1942 but were not permitted to publish their findings until April, 1945.

By that time the American work, which had focused particularly on the selective herbicidal action of 2,4-D, had already been published and its use as a herbicide patented. By 1946 more than 2000 tonnes of 2,4-D were being used. Its production continued to rise rapidly and many related compounds were also sold commercially, although none more effective than 2,4-D, MCPA and 2,4,5-T (trichlorophenoxyacetic acid) were found.

Darwin's curiosity, and that of his successors, had led indirectly but inexorably to one of the most powerful and pervasive innovations in the whole of agriculture. 2,4-D and related compounds made possible the selective and effective control of weeds, raising crops yields, simplifying crop rotations, reducing energy use in crop production and opening the way to the development of minimum tillage techniques.

## 7.7 DDT: the insecticide revolution

The discovery of the remarkable insecticidal powers of *p,p'*-dichlorodiphenyltrichloroethane (DDT) occurred just as World War II was beginning. Paul Müller (1899–1965) had been commissioned in 1935 by J.R. Geigy AG of Basel, Switzerland to find an effective synthetic insecticide for the control of agricultural pests. He accomplished this in September 1939 when he synthesized DDT (Figure 23) and tested it on blowflies.

Müller had synthesized and tested hundreds of substances before DDT and had noticed that compounds with a $-CH_2Cl$ group and with a *p,p'*-dichlorodiphenyl structure often showed considerable oral toxicity for moths. His tests of DDT toxicity for flies were nearly ignored because, even after thorough cleaning of his cages, untreated flies fell to the floor on touching the walls. Such rapid knockdown had never been associated with prolonged residual toxicity among earlier insecticides, such as pyrethrum and rotenone.

Further syntheses of related compounds by Müller did not reveal any with an effectiveness on insects comparable with DDT or with such a lack of human toxicity. Meanwhile, the range of its usefulness was extended beyond agriculture to pests of stored grain, timber and woollen textiles, and to vectors of human diseases, such as the lice that transmit typhus and the mosquitoes that transmit malaria.

Clearly, DDT could be of great military significance in 1942 and, being rigorously neutral, the Swiss government advised both sides in the war of its potency as an insecticide. Its production was given a high priority in Britain, and it was later used by the Allies to control typhus epidemics in Italy and the Rhine at the end of the war and to control malaria in south-east Asia. Its success stimulated the search for other synthetic organic pesticides, leading to the introduction of 25 additional ones between 1945 and 1953, including chlordane, toxaphene, aldrin, dieldrin, endrin, heptachlor and parathion. Müller was awarded a Nobel

Prize in Medicine in 1948 for the search begun in 1935, achieved by 1939 and brought to the Swiss market by 1941. Such speed in the approval of agriculturally useful compounds is no longer possible, nor would such freedom for individual research be common today.

In his Nobel Lecture Müller[144] listed the properties of his ideal insecticide as follows: 1. Great insect toxicity; 2. Rapid onset of toxic action; 3. Little or no mammalian or plant toxicity; 4. No irritant effect ...; 5. Wide range of action ...; 6. Long persistent action ...; 7. Low price. DDT met all these criteria and was quickly recognized as a revolutionary development in pest control, both for agriculture and for medicine. The pesticide revolution led to widespread reliance on insecticides for pest control, often with heavy and multiple spraying by the calendar rather than by the need. The unfortunate consequence was that, by the 1960s, desiderata 5 and 6 in Müller's list were being questioned because of their association with the loss of many beneficial insects, and with accumulation of DDT in the food chain.

DDT production reached its peak in the USA in 1963, when 80,000 tonnes of it were produced. Its first agricultural success had been in Switzerland where it controlled the Colorado potato beetle and safeguarded Swiss potato production during and after World War II. Elsewhere, military requirements for nearly all the DDT being produced delayed its use in agriculture until after the war. Then, however, it had a wide-ranging impact on the protection of many field crops, fruit trees, vines, vegetables and flowers.

Earlier insecticides had been not only far less effective but also poisonous to people (e.g. nicotine and the arsenicals), of limited availability (e.g. pyrethrum) or unstable in sunlight or air. DDT had none of these limitations, but its cheapness, availability and stability eventually hastened its demise. On cotton crops it was initially highly effective against the boll weevil and the boll worms, but it was used so frequently and abundantly over such huge areas that resistant forms soon appeared, as they did eventually in more than two hundred other insect pests, including house flies, mosquitoes and lice.

Nevertheless, there are still many insect pests in which resistant lines have not developed and, with more precautionary use, DDT could still be a remarkably safe, cheap and effective insecticide. There are now legal limitations on its use in many developed countries, but that is another story, one which Kenneth Mellanby[133] believes should be reconsidered.

## 7.8 Tissue culture: Trojan horse for things to come

'I went into science myself a great deal at one time, but I saw it would not do. It leads to everything; you can let nothing alone.' So says Dorothea Brooke's loquacious uncle in George Eliot's *Middlemarch*. The artificial culturing of plant tissues seemed likely to be far removed from the needs of agriculture, yet it illustrates Mr Brooke's assertion.

Gottlieb Haberlandt was the first to consider the purposes and potential of plant tissue culture, in Berlin in 1902, but failed in his own attempts to culture mature grass leaf tissue. His objective had been to show that plant cells from any organ or tissue retained the capacity to form a whole new plant. Understanding, not application, was his aim.

It was not until 1934 that two rivals in the search, Philip White[223] of the USA and Roger Gautheret[70] of France, both succeeded in getting the survival of plant tissues in culture, using tomato roots and willow cambium respectively. Many experiments to improve the nutrient medium for tissue cultures followed. In 1937 White discovered the beneficial role of vitamin $B_1$ for root growth while Gautheret found the enhancing effect of auxin on tissue growth. Then in 1939, on the verge of World War II, they both satisfied the criteria for successful plant tissue culture of sustained and undifferentiated growth to form callus, Gautheret with carrot roots, White with tobacco stems.

Reviewing the field in 1956, White foresaw its 'applications' in terms of better understanding of the nutrition, metabolism, hormonal relations, genetics and responses to diseases of cells, and of how they interact to form organs. Clearly tissue culture was still a pure science.

The next major discovery was of the plant growth substance kinetin, to which cell division is highly responsive. Its addition greatly improved growth media, and Skoog and Miller found in 1957 that callus cultures would form either roots or shoots depending on the auxin/kinetin ratio. The stage was set for controlled organ formation and the regeneration of whole plants from tissues. This was first achieved with carrots in 1958, but it took much longer with many cereals and legumes. The regeneration of whole plants from single cells was first achieved in 1965, with tobacco.

In 1956 workers in Japan used tissue culture to rescue immature rice embryos. By 1959 haploid tissues (with only half the usual complement of chromosomes) had been successfully cultured, as were shoot apices with the aid of gibberellins. By the end of the 1950s, therefore, several practical applications of tissue culture not foreseen by White in 1956 were already opening up. Embryo rescue techniques would obviously be of great value to plant breeders wishing to make 'wide crosses' in order to introduce useful characteristics from wild or distant relatives, and so it has proved. Meristem culture became particularly valuable for the eradication of viruses from vegetatively propagated plants, such as root crops, strawberries, asparagus and orchids.

The 1960s introduced further advances and opportunities, often unforeseen. In 1960 Cocking, in England, succeeded in removing the cell walls in tomato roots with cellulase to isolate naked but viable protoplasts. This opened the way, after much further research, for the uptake by the protoplasts of cell organelles, viruses or alien genetic material to produce genetically modified cells before the cell walls reformed and plantlets were regenerated. Tissue culture became the Trojan horse for the introduction of new genetic constructs into old crops, by a variety of means, and cultured cells could be screened for antibiotic or herbicide resistance markers to identify which cells had been transformed.

The 1960s also saw the introduction of anther culture by Guha and Maheshwari of India. Their aim had been the study of cell division but they noticed the many embryo-like structures developing inside their tissue-cultured anthers. The haploid plantlets from the germinating pollen grains gave immediate access to large populations of genetically uniform plants, which until then had required many years of laborious backcrossing. Anther culture was immediately put to use in the breeding of rice in Japan and then on a huge scale with many crops in China in the 1970s.

During the 1970s the ability to store plant tissue cultures by freeze preservation was developed. This led to an active interest by gene banks in the use of this technique for the conservation of genetic diversity, particularly of vegetatively-reproduced crops where long-term conservation has been a major problem.

As the techniques of tissue and cell culture continue to develop, new opportunities are recognized and new directions taken. More recently, for example, there has been much emphasis on increasing the production of 'secondary metabolites' such as alkaloids, flavonoids, cardiac glycosides, antibiotics and volatile oils in tissue cultures, and their use in fermentation systems.

But enough has surely been said to illustrate how one area of what was originally basic research has opened up a continuing stream of opportunities for crop improvement and agricultural benefit. These have evolved from experiments in many different countries, and the synergisms between their findings have led to frequently unforeseen benefits in many contexts. Truly, as Mr Brooke claimed, science 'leads to everything; you can let nothing alone'.

CHAPTER 8

# The fourth billion (1960–1975)

## 8.1 Introduction: focus on development

It took a century to add the second billion to the world's population, and 33 years to add the third, but the fourth took only 15 years. The highest growth rate of the world population, of 2.1% per year, was reached in the 1965–70 period. Not long after our numbers reached three billion and the technological triumph of sending satellites into orbit, we saw the first photographs of the earth from space, a small planet in need of care and maintenance if humankind was to survive. The growing concern for 'only one earth' became the focus of the first United Nations Conference on the Human Environment in Stockholm in 1972 as the world population approached four billion with what Barbara Ward and René Dubos referred to as 'sober optimism'.

However, the contrast between the rapid rise in population and the sluggish increase in food production in the less developed countries, where most of the fourth billion were added, 69% of them in Asia, demanded attention. Both the Rockefeller and the Ford Foundations had sent missions to India in the late 1950s to examine how cereal yields there could be raised. The mid–1960s droughts in India exacerbated concerns, as exemplified by the conclusion of the 1967 report of the US President's Science Advisory Committee that 'The scale, severity and duration of the world food problem are so great that a massive, long-range innovative effort unprecedented in human history will be required to master it.'[164]

The following year saw the publication of the Paddock brothers' book predicting *Famine–1975!*, engendered by the turn in the tide of grain flow between the more and the less developed countries. That year also saw the establishment of the Club of Rome to examine 'the predicament of mankind', which led to the publication of *The Limits to Growth*. This was based on the recently-introduced and plausible art of dynamic modelling on a global scale, but for all its apparently sophisticated quantitative approach, the exponential assumptions inevitably led to predictions of disaster. Yet another strand in the prevailing pessimism was the sharp decline in carryover stocks of grain in the world, from about 100 days' supply in 1960 to only about 40 days in 1974 following the oil price crisis of 1972/3, which led to a temporary slump in grain production.

In 1971 the Consultative Group on International Agricultural Research (CGIAR) was established to support a more comprehensive system of international agricultural research centres to work in partnership with the less developed countries. Its founding fathers had been greatly encouraged by the early successes of the dwarf wheat and rice varieties first released in 1962 and 1966 respectively. As early as 1968 the USAID administrator W.S. Gaud had referred to their impact as a 'Green Revolution', and so it proved to be.

By 1970 the dwarf varieties occupied almost a quarter of the total wheat area in developing countries (excluding China), 40% by 1975 and today about 70%. Although the initial adoption of the dwarf wheats was mostly on irrigated land, they soon spread to the rainfed areas. Because they did not lodge (i.e. fall over) so readily and because they had a higher yield potential, their spread was accompanied by increasing use of fertilizers. The dwarf wheats therefore acted as a spur to the introduction of agronomic advances into developing countries, leading to a rapid increase in yields and production.

The dwarf rice varieties had a similar course of uptake and impact. By 1980 they occupied about 40% of the area sown to rice in south and south-east Asia, and 74% by 1990. As with the dwarf wheats their use was often associated with irrigation and greater use of fertilizers. Together these resulted in a steady annual increase in rice yield and production of about 3%, to which the new varieties, irrigation and fertilizers contributed about equally. Thus, as a result of the synergistic interactions between these three factors there was indeed a Green Revolution, and the supply of staple foods more than kept pace with the rising population. Indeed, from 1975 on the FAO index of food supply per head in developing countries began to rise, and the real prices of wheat and rice continued to fall (Figure 24).

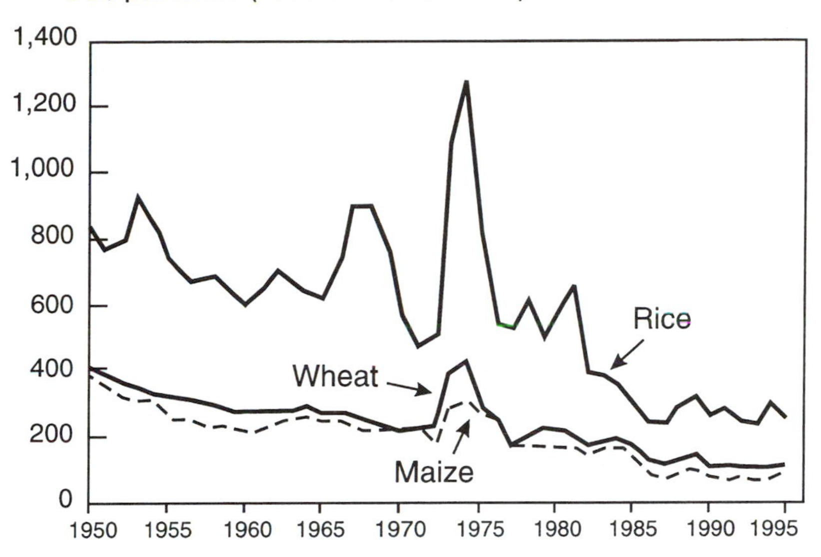


**Figure 24** Trends in real world prices (in $US 1990) for wheat, rice and maize. (Data courtesy of M.W. Rosegrant.)

As often, however, success spawned reactions. Some human nutritionists, most notably the Protein Advisory Group of the United Nations Organization, became concerned that the very success of the dwarf cereals, partly at the expense of the grain legumes, would exacerbate a perceived 'protein gap' in the diets of developing countries. It turned out, however, that calorie supply was, in most cases, more limiting than dietary protein. Moreover, at least in India, the dwarf cereals not only increased the supply of protein, but substantially reduced its price.

During the 1970s many social scientists also criticized the Green Revolution on a variety of other grounds: according to them it favoured the larger farmer; it reduced employment opportunities, particularly for women and landless labourers; it led to both national and individual dependence on agrichemical companies and creditors; it affected the health of both farmhands and rural environments adversely; it disadvantaged the grain legumes and weakened crop rotations; it made yields more variable and genetic resources more vulnerable, and so on.

The criticisms were salutary but overly pessimistic in most cases. There had to be costs – social, environmental and agricultural – if the world food supply was to be increased rapidly. However, rural employment opportunities increased, small farmers eventually benefitted as much as their larger neighbours, the 'revolution' spread far beyond the favourable irrigated areas, and food supplies did not become more vulnerable[90]. The ultra-poor and hungry remain so, unfortunately, but we should not expect agricultural progress to stand proxy for social reform. As Lipton and Longhurst[119] put it: 'If social scientist(s) had designed a blueprint in 1950 for pro-poor innovation, it would be like the MVs (modern varieties).'

By raising yields substantially the dwarf cereals took pressure off the expansion of arable area to such an extent that further land clearing for agriculture merely matched the losses to urbanization and soil degradation. However, the area under crops still increased to some extent through the reduction of fallow land and more multiple cropping. The increases in cereal yields also made it clear that agricultural intensification can succeed at low latitudes in spite of their climatic and biotic limitations. Possibly associated with these transforming effects of the dwarf cereals, lending for agriculture and rural development by the World Bank rose from about 1% in 1959 to almost 40% of its loans 20 years later. The unfashionable view put forward by T.W. Schultz[186], that the transformation of traditional agriculture depends on and justifies investment in agriculture, now commanded attention.

The growing of dwarf cereal varieties also spread through the developed countries as the world population approached four billion. Their resistance to lodging after heavy dressings of nitrogenous fertilizers was the key to further intensification of agriculture, made worthwhile by the rise in yield potential as the harvest index increased (see p. 140). But with intensification came problems. Surpluses, surfeit and subsidies troubled the 30% of people living in the developed countries while the less developed 70% faced deficits, debts and deficiencies.

Along with their surpluses of grain production, the developed countries also began to register the social and environmental perils of excessive or careless applications of agrichemicals. In 1962 this growing concern was crystallized by Rachel Carson's book *Silent Spring*, which focused on the environmental hazards of pesticides such as DDT, herbicides and other chemicals used in agriculture. Ironically, the very property of DDT which had attracted the attention of its discoverer, Paul Müller, namely its residual toxicity, now became its undoing.

Unwelcome side-effects of many agrichemicals were uncovered, including those of fertilizers, such as eutrophication and nitrate contamination of groundwaters. But research to ameliorate these problems led to compounds which were more specific in their targets, with higher biological activity or reduced persistence, as well as to slower release, better placement and more informed timing of applications. Techniques for applying irrigation water were greatly improved, as were the strategies for using resistance genes in plant breeding programmes to minimize the use of agrichemicals. Rachel Carson and her successors had a salutary effect on agricultural research, and there was an air of confidence as we approached our four billion, exemplified by the rash declaration of the 1974 World Food Conference 'that within a decade no child will go to bed hungry, that no family will fear for its next day's bread and that no human being's future and capacities will be stunted by malnutrition', despite the famines in the Sahel in 1972–4 and in Bangladesh in 1974.

## 8.2 The dwarfing of wheat and rice

The greatest impact on world food production as the population grew towards four billion came from the deployment of the dwarfing genes in wheat and rice in the 1960s. Like many other advances we have considered, its origins were much earlier and its impact is still increasing, but it came to fruition just when the escalating use of nitrogen fertilizers made it necessary and the development of herbicides made it possible.

Until the 1960s, tallness in cereals was an advantage. The straw was valuable for roofing, mulching and for livestock feed and bedding, yielding large amounts of farmyard manure, important as fertilizer and a valuable fuel in many developing countries. Height was also needed to allow the cereals to compete more effectively with weeds. With rice, particularly, taller plants were better adapted to traditional harvesting by hand.

In several east Asian countries in the latter half of the 19th century, however, shorter varieties of both wheat and rice began to be used. Less dependence on weed-infested farmyard manures and more intensive hand weeding reduced the need for tallness, while their heavy use of fertilizers made shorter, stronger stems desirable to avoid the lodging of crops. When Horace Capron, the US Commissioner of Agriculture, visited Japan in 1873 he wrote: 'The Japanese farmers have brought the art of dwarfing to perfection.' He noted that the wheat ears were heavy but borne on such short stems 'that no matter how much manure is used ... on the richest soils and with the heaviest of yields, the wheat stalks never fall down and lodge.'[49]

Similar dwarf wheats were known in Europe at that time. In England in 1847 'Piper's Thickset' wheat was recommended for use on rich soils and in France the dwarf 'Blé Précoce du Japon' was sold commercially from 1882. After about 1900 the height of many European wheat varieties began to fall from one and a half metres to reach about one metre by 1960. However, this change was gradual and did not involve the use of dwarfing genes of major effect. For these we have to return to east Asia.

In Japan in 1917 the semi-dwarf wheat variety Daruma – which may have come from Korea – was crossed with an American variety and the progeny of that cross were then crossed with another American variety in 1925. Selection from the progeny resulted in a semi-dwarf variety called Norin 10 which was included among the dwarf Japanese varieties taken back to the USA by S.C. Salmon in 1946. There it was crossed by Orville Vogel with an American variety, Brevor, to give rise to the variety Gaines which was suitable for use with heavier applications of nitrogen fertilizers. Seeds from the Norin 10 × Brevor cross were also sent to Norman Borlaug in 1954, to introduce dwarfness into the Mexican wheats, with almost immediate success, leading to the release of varieties Pitic and Penjamo in 1962 and Sonora in 1963. These varieties and their many successors spread like wildfire through the developing world and now occupy about 70% of its total wheat area.

The breeding of dwarf rice for the subtropics followed a similarly prolonged and fortuitous route. As with wheat in Japan in the late 19th century, so also with rice there was emphasis on the breeding of shorter varieties which yielded well in response to fertilizer applications. Subsequently, in the 1920s, Japan established a breeding programme in Taiwan for short, fertilizer-responsive *japonica* rices known as ponlai varieties which soon became dominant there. When Chinese rule was restored to Taiwan in 1945, crosses among local *indica* varieties included the dwarf Dee-geo-woo-gen from which the outstanding local variety Taichung Native 1 was selected. When the International Rice Research Institute (IRRI) began its breeding programme in 1962, Dee-geo-woo-gen was crossed with the tall Indonesian variety Peta and a dwarf selection from this (the eighth) cross was released in 1966 as IR8.

Lodging-resistant, fertilizer-responsive rices for the more tropical regions were born, and spread rapidly.

In rice there is only one widely-used dwarfing gene, which is recessive and inhibits production of the elongation hormone, gibberellic acid. In wheat there are more than ten 'reduced height' (Rht) genes, of which only three have been widely used. The two most widely used dwarfing genes, Rht1 and Rht2, came from Norin 10. Instead of blocking gibberellin production they confer insensitivity to it, whereas Rht8, which came from the Japanese wheat variety Akakomugi, blocks gibberellin synthesis. The Rht8 gene has long been used in Italian wheat breeding and varieties with it seem to be well adapted to East European conditions, whereas Rht1 and/or 2 have been used in all the 'Mexican' wheats and in breeding programmes throughout the Americas, Western Europe and Asia. So there may still be scope for a better-informed deployment of these genes.

The dwarfing genes were introduced into rice and wheat to reduce lodging at higher rates of fertilizer use, i.e. to reduce yield losses. Subsequently, however, they have had an even greater, but largely unanticipated, impact on world food production by increasing the yield potential of wheat and rice, as we shall see in the next section. Both chance and design have played a part in their discovery and in their exploitation.

## 8.3 The rise of the harvest index

The headlines of the late 1960s often referred to the transformation in world food supplies effected by 'miracle rice' and 'the wonder wheats' without specifying in what way they were miraculous or wonderful. Admittedly they did not fall over and lodge after heavy dressings of nitrogen fertilizers, but more than that was implied. They did not germinate sooner or grow faster, nor was their photosynthesis more efficient as sometimes suggested, so to what did they owe their superiority?

In 1962 a Dutch crop scientist, W.H. van Dobben, had shown that the substantial increase in yield of Dutch wheat varieties released between 1902 and 1961 was not accompanied by any increase in total crop weight but by a rise in the proportion of it in the grain, and therefore in yield. In 1920 the English crop scientist E.S. Beaven had called this proportion the 'migration coefficient' and had described the last stage of cereal crops as follows: 'Some time ... before the grain is ripe the plant ceases to gain in weight ... Its last effort is to transfer its accumulated reserves into the grain. But all, or very nearly all, the dry matter of the grain is first stored up in the leaves or stems of the plant ... It is mainly ... on the extent to which this 'uplift' takes place that plenty or scarcity of the staple food of man depends.'

That view of crop development may have been correct for the varieties and agronomy of the 1920s. But as the use of nitrogen fertilizers grew, and particularly if irrigation was available, varieties whose upper leaves remained green and active after flowering were selected. Even by 1945 it was being suggested that most of the sugars required for grain growth came from 'current account' rather than from 'savings'. In good wheat crops these days the reserves stored in the stem before flowering may contribute only 5% of grain weight, although rather more in rice.

The proportion of total plant weight that ends up in the grain is now called the 'harvest index'. For British wheat varieties in the 1920s the harvest index was about 35%; for modern varieties of both wheat and rice it is commonly 50–55%, having risen steeply since the dwarfing genes were introduced (Figure 25). Stem weight in the cereals is roughly proportional to stem length, so the great reduction in the height of wheat and rice crops released a lot of dry matter for investment in organs other than the stem. Initially much of this went into the growth of additional tillers (branches), more than were needed, but selection has gradually shifted much of it into additional grains, thereby raising the yield potential.

Thus, the fertilizers that made stem shortening necessary, and the herbicides that made it possible for dwarf varieties to be grown successfully, led to a shift in the investment of resources in wheat and rice plants from stem to grain. By extending the photosynthetically active life span of leaves, especially in combination with irrigation, fewer leaves and tillers were needed to ensure a crop. Better protection of crops against pests and diseases reduced the need for reserves to aid recovery from defoliation. And improved supplies of nutrients and water meant that a smaller root system would suffice. Thus greater agronomic support for crops through irrigation, fertilizers, herbicides and pesticides has, beyond the directly beneficial effects, allowed plant breeders to select for greater investment in the grain. The rise in the harvest index of wheat, rice and other cereals since the 1960s has, there-

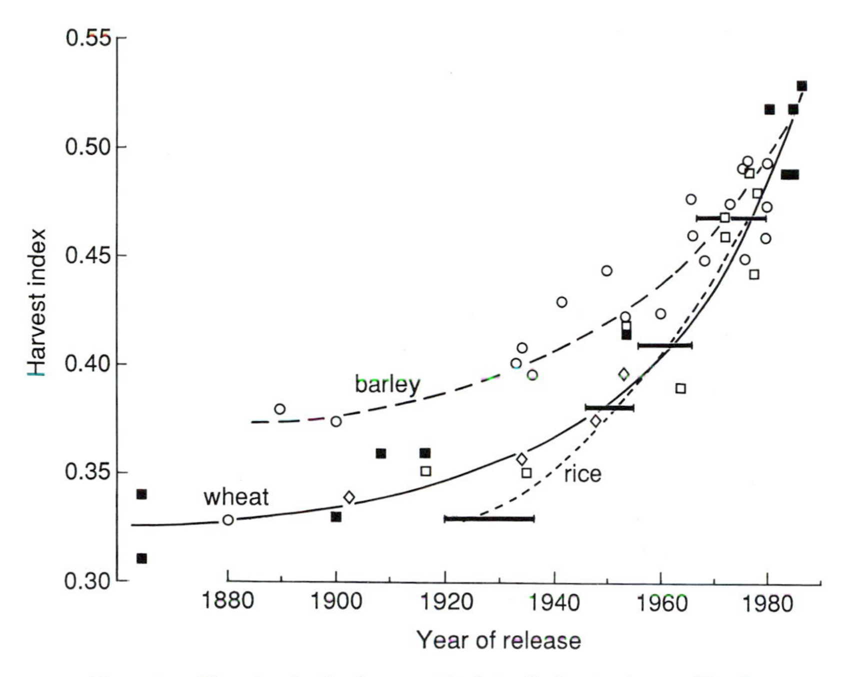


**Figure 25** The rise in the harvest index of wheat, rice and barley varieties with the year of their release[58].

fore, been an indirect consequence of the intensification of agriculture.

Clearly there are limits to how much further that rise – the source of recent increases in yield potential – can go. The harvest index depends also on the life span and growth habit of the crop. Although further dwarfing, e.g. with the more extreme Rht3 gene in wheat, is possible, it is likely that the harvest index of wheat and rice will not increase much beyond 60%. If so, we are approaching the end of the era of increasing yield potential via the harvest index, while remaining unsure of what other avenues can be exploited.

## 8.4 Silent Spring: the gathering storm

John Keats had the anguish of a bewitched lover in mind when he wrote 'The sedge is withered from the lake, and no birds sing', but in *Silent Spring* Rachel Carson[34] wrote of the alienation between modern agriculture and nature: 'Single crop farming does not take advantage of the principles by which nature works ... Nature has introduced great variety into the landscape, but man has displayed a passion for simplifying it. Thus he undoes the built-in checks and balances by which nature holds the species within bounds.'

Carson's primary focus was on the harm being done to wildlife, and possibly to human health, by the excessive and indiscriminate use of 'the miracle insecticide' DDT in agriculture, forestry and recreational and public health programmes. Many biologists and naturalists had expressed similar concerns and had gathered evidence before her. Indeed, *Silent Spring* was based on their work, as Carson makes plain. As early as 1945 Kenneth Mellanby had said that field research on DDT was needed 'on all manner of apparently unimportant insects and other forms of life, to ensure that there was no serious effect on the 'balance of nature' with subsequent disastrous effects.'[133]

What Rachel Carson achieved was to bring the accumulating evidence of the harmful side-effects of modern insecticides together in a comprehensive and persuasive synthesis and, gently but insistently, to inject a sense of urgency into the issues. *Silent Spring* is an unusual but effective combination of careful analysis, biological explanation, literary elegance and a call to arms. It was an instant success. A recent survey of the ten most influential books of the 20th century selected *Silent Spring* for the 1960s, ranking Carson alongside such authors as Keynes, Freud and Orwell.

'How to simplify without error' was her constant concern as she battled 'the myth of the harmlessness of DDT'. She acknowledges early on that other insecticides such as chlordane, heptachlor, dieldrin, aldrin, endrin and the organophosphates are more dangerous. But DDT had been around the longest, used in greatest amounts for the most varied purposes and was the most persistent, hence her focus on it. Carson pursues it through many surface and underground waters, such as Clear Lake in California where many western grebes had died after the lake had been sprayed to rid it of the gnats which annoyed anglers. Clear Lake provided a telling example of the progressive concentration of DDT up the food chain, from 0.02 ppm in the water, through 5 ppm in plankton, 40–300 ppm in plant-eating fish, to up to 2500 ppm in carnivorous species.

Carson then considers accumulation in the soil before turning to various effects of 2,4-D and other herbicides on vegetation and wildlife. Her most effective chapter is that on the loss of birds, which gave the title to her book. The ramifying effects of the DDT spraying campaigns against Dutch elm disease provided a powerful example of the 'web of life', as did the decline of the eagles after 1947.

Although the greatest amounts of DDT were used in agriculture, which made modern agriculture a prime culprit in *Silent Spring*, Carson's examples of misuse are mostly drawn from forestry and public health and nuisance spraying programmes, which were not only dangerous but ineffective. One of her relatively few agricultural examples concerned the loss of game birds in England in 1960/61, due to the presowing treatment of seeds with insecticides and mercurial dressings, largely solved by confining the use of such dressings to autumn-sown crops. In her final chapter, she urged the use of 'the road less traveled', of a more biological, less chemically-oriented approach to the control of agricultural pests and public health problems. She adds, however, 'It is not my contention that chemical insecticides must never be used.'

*Silent Spring* unleashed fierce public debate on the issues Carson had raised and, in the USA, extensive litigation by citizen action groups, which has been extensively analysed from several points of view. The ultimate outcome was the decision by William Ruckelshaus, administrator of the Environmental Protection Agency, to ban all uses of DDT except those related to essential public health purposes. Within 30 years of its first use, therefore, the 'miracle insecticide' was banished from agriculture in most developed countries because of two of the characteristics which its discoverer had listed as essential in his Nobel lecture: wide range of action and long persistence.

## 8.5 The protein gap and high lysine maize

Just when dwarf wheat and rice began to have a favourable impact on world food production, another cause for alarm was raised and blamed on the cereals, namely 'the protein gap' and 'the impending protein crisis'.

The United Nations Organization had established a Protein Advisory Group (PAG) in 1955 which, as the dwarf cereals spread rapidly and displaced legume crops in the 1960s, expressed concern that calorie malnutrition would soon be replaced by protein malnutrition. With the best of intentions they had conflated the West African condition of kwashiorkor, which may be due to insufficient protein in the diet, with the much more common marasmus associated with inadequate dietary energy intake, to conclude that protein malnutrition was widespread.

Early estimates of human energy and protein requirements, extending back to a League of Nations commission in 1936 which advocated a daily intake of 1 g protein per kg adult body weight, were revised by expert committees of FAO and WHO (the World Health Organization) in 1957, 1965, 1973 and 1985[33]. In the early sixties, when PAG was raising the alarm, the estimated protein requirements for most developing regions greatly exceeded actual supplies, as estimated from FAO's food balance sheets, far more so than did the energy requirements. PAG estimated that more than a third of the world population was not getting enough protein. So great was the apparent shortfall, and so alarming the suggestion that the dwarf cereals would exacerbate it, that a spokesman for FAO declared that 'deficiency of protein in the diet is the most serious and widespread problem in the world' and a United Nations report in 1968 referred to 'the impending protein crisis'.

However, these concerns diminished in 1973 when the FAO/WHO Expert Committee reconsidered our energy and protein requirements. While those for energy were modified only slightly, the protein requirements were substantially reduced. The protein gap disappeared overnight, except for about 5% of the world's malnourished people, particularly those subsisting on starchy foods such as cassava, yams and plantain. McLaren[125] called it 'the great protein fiasco', while Waterlow & Payne[216] announced 'the protein gap is a myth'. This view is now widely accepted and has influenced policies for aid.

The PAG now agrees that protein supplies available for human consumption are ample for the world as a whole, but that because of poor distribution and the greater protein requirements of young children, pregnant and lactating mothers and the sick, the protein problem remains. They agree that this is more a problem of inadequate purchasing power and knowledge, and it was here that the dwarf cereals actually helped. In India, for example, the spread of high-yielding varieties increased the availability of wheat and reduced the real price of protein by one third between 1965 and 1977.

As well as the concerns of PAG about the *amount* of dietary protein, there were others about its nutritional *quality* as determined by amino acid composition. These focused particularly on maize, in which the main storage protein, zein, was known to be very low in the essential amino acid lysine, thereby greatly reducing the nutritional value of maize protein.

In 1964 three American scientists found that the *opaque*–2 mutant of maize largely suppressed the production of zein, one consequence of this being a rise in lysine content from 1.6 to 3.7% of the seed protein. The content of several other amino acids, such as tryptophan, arginine and aspartic acid also increased. Another consequence, however, was that grain size and yield were substantially reduced, but it was expected that empirical selection would eventually lead to a compensating increase in other seed proteins and the restoration of yield.

In its report to the United Nations in 1968, PAG had recommended the prompt development of high lysine maize varieties and, with funding from the United Nations Development Program, the International Centre for Maize and Wheat Improvement (CIMMYT) in Mexico took up this challenge in 1970. The protein quality of a range of tropically-adapted maize varieties was improved substantially. The accompanying soft chalky character of the kernels was corrected, as was the vulnerability of high lysine maize ears to storage pests and diseases. Finally, yield levels comparable to those of the best varieties were achieved.

And the result? Remarkably little interest among developing countries in growing high lysine maize, to the extent that CIMMYT had to discontinue further work on a programme which had accomplished all the objectives set for it. In the fifteen or so years required to achieve these objectives the importance of the content and quality of proteins to the world food problem had been downplayed. Most of the poor who depended on maize combined it with beans or other pulses in their diet, thereby balancing their amino acid intake without a need for high lysine maize. The feed industry could more readily use industrial lysine supplements. Without a price premium for high lysine maize, farmers were loath to grow it, yet the potential beneficiaries would be those least able to afford such a premium. There is a lesson here for those who call for plant breeding programmes to raise the mineral and vitamin contents of staple crops where varietal turnover is rapid, as well as for well-intentioned scientists who seek technological solutions to problems of social and economic inequity.

## 8.6 Latitude and the Green Revolution

All our major crop plants originated in latitudes lower than those now occupied by the more developed countries. Despite their origins, however, and despite the fact that total annual growth is greater at low latitudes, grain yields are highest in the more developed countries. Average maize yields at the higher latitudes are about four times higher than those in the tropics, and even national average rice yields are three times higher. To what extent does this gap reflect economic development as against environmental differences?

The higher temperatures at lower latitudes, particularly the warmer nights, have a major depressing effect on yield per crop in two ways. Losses by respiration rise rapidly as the temperature rises, but even more important is the shortening of the period of growth before flowering and of the duration of grain growth after it. This factor alone greatly reduces potential crop yields in the tropics. Heavy cloud cover in monsoonal areas, with lower light intensities, limits wet season rice yields. In the dry season, on the other hand, the effectiveness of high solar radiation is limited by the much shorter daylengths of the tropics compared with the higher latitudes. Indeed, one study of maize yields by latitude found them to be most closely correlated with natural daylength[38]. On top of these effects come the generally greater losses from pests and diseases in the lower latitudes, partly because there is no closed season to limit their multiplication and partly because at least some of the pests and diseases were left behind as the crops were introduced to higher latitudes.

All these unavoidable effects substantially reduce yield per crop at lower latitudes, but reducing them still further until the 1960s was the limited economic development of many low latitude countries. As one measure of this consider their use of fertilizers. In 1961 the average use by developing countries of nitrogen, phosphorus and potassium fertilizers was only 6 kg per hectare, compared with 45 kg $ha^{-1}$ in developed countries. Thirty years later the developing countries used 82 kg $ha^{-1}$ compared with 116 in the developed countries, well on the way to closing the gap.

As we have seen, the greater use of fertilizers was made worthwhile by the development of the dwarf cereals. Similarly, the spread of the dwarfs and the greater use of fertilizers was encouraged by the increasing extent of irrigation. Other inputs, such as herbicides, were also synergistic. The increases in agricultural productivity in turn supported other facets of economic development.

Nevertheless, given the climatic limitations on yields per crop at low latitudes, they are unlikely ever to match those of the more developed countries at high latitudes and should not be expected to do so. Moreover, the warmer temperatures which limit crop duration and yield confer a compensating advantage by allowing three or even four crops to be harvested each year, particularly where irrigation is available for dry season cropping. Selection for earliness can enhance this advantage and in the Philippines four successive rice crops within 335 days have yielded 25.7 tonnes of grain $ha^{-1}$, rather more than the world record maize yield from one crop at high latitude of 23.2 t $ha^{-1}$.

Although the 'Green Revolution' is often viewed as a re-run of the previous agricultural revolutions at higher latitudes, there were several quite different elements in it, such as the greater scope for multiple cropping. Like the earlier revolutions it acted as an 'engine of change' for the wider economy but engendered adverse effects on some groups. These were seized on, almost avidly, by many social scientists in the 1970s, to denigrate the value of the Green Revolution. Some of their apprehensions have been shown since to be unfounded, while others have been acted upon. Concerns about the environmental effects of fertilizers, pesticides and irrigation remain, but are common to all agricultural intensification.

Despite it being damned by many social scientists, the farmers of the developing countries voted for the Green Revolution with their fields, and their cereal yields began to rise in the late 1960s, keeping pace with population growth until 1975 when our numbers reached four billion. The linkages between latitude and yield were being weakened.

## 8.7 The internationalization of agricultural research

It all began when Henry A. Wallace, erstwhile promoter of hybrid maize and then Vice-President of the United States, was on a state visit to Mexico in 1940. After his return, he persuaded the Rockefeller Foundation that their programme on the improvement of public health in Mexico should go hand in hand with another on the improvement of agriculture. This led to the Foundation initiating maize and wheat breeding programmes in Mexico, and to Norman Borlaug arriving there in 1944 and breeding the semi-dwarf wheat varieties which swept through the developing countries 20 years later.

In the early 1950s the Foundation was also considering how it could help to improve food prospects in Asia. Warren Weaver and George Harrar proposed the establishment of an International Rice Research Institute (IRRI) to investigate the problems of rice crops 'independent of geography, and ... certainly independent of political boundaries.'[8]

This proposal was shelved until 1958 when the Ford Foundation indicated its willingness to share the cost of the Institute with the Rockefeller Foundation on the grounds that: 'At best, the world food outlook for the decades ahead is grave; at worst, it is frightening.' IRRI, the first of the growing line of international agricultural research centres, was established at Los Baños in the Philippines, in 1960. The two foundations also transformed the Mexican programme into an International Centre for Maize and Wheat Improvement (CIMMYT) at El Batan in 1966, just as the widespread adoption of the dwarf wheat and rice varieties by many developing countries took off.

They then established two more centres, one (IITA, Nigeria, 1967) to seek viable alternatives to the bush fallow systems still used throughout the humid tropics, the other (CIAT, Colombia, 1968) to develop better farming systems for the lowland tropics. But by mid-1968 the two foundations realized that their resources were stretched to maintain these four international centres and that it was 'time to go public'.

Other foundations and aid agencies soon indicated a willingness to help and a series of meetings to consider how this might be done was held at the Rockefeller conference centre in Bellagio, Italy. In opening the first of these, in 1969, the chairman referred to the early impact of the dwarf wheat and rice varieties as follows: 'These results generate optimism regarding the possibility of increasing food production rapidly enough to close the food gap over at least the next two or three decades, and thus buying time for population programs to reduce the rapid rate of population growth.'[8] Robert McNamara, the President of the World Bank, suggested the formation of a consultative group of donors to support the expansion of a system of international agricultural research centres, several more of which were already under consideration. Eventually this is what happened, with the Consultative

Group on International Agricultural Research (CGIAR) being established in May 1971 under the co-sponsorship of the World Bank, FAO and the United Nations Development Program.

Following the establishment of the CGIAR, the number of international centres expanded to ten by 1974, to 13 by 1980, and then to 16. They are distributed throughout the developing world plus the Hague, Rome and Washington. Each has an independent and international Board of Trustees, and there is a considerable diversity of approach among them. Some are like the first two highly successful centres in focusing on the problems of one or a few crops. Others are like IITA and CIAT, with a broader eco-regional focus such as the semi-arid tropics or the cooler dry areas. Some concentrate on individual factors affecting production, such as irrigation or animal diseases, while others have a more disciplinary orientation, such as forestry or aquatic resource management. Although most of the crop improvement centres maintain their own 'gene banks', one institute is dedicated to overall strategies for the conservation of genetic resources. Another, IFPRI, analyses policies and strategies for meeting the food needs of the developing world on a sustainable basis.

This informal, genuinely international and highly effective 'system' of agricultural research centres has evolved over the years, becoming globally more integrated. Although its total expenditure represents only a few percent of the overall cost of agricultural research in the developing countries – much of which is also borne by aid agencies – it has played a catalytic role in transforming attitudes to agricultural research in them. Through its networks and training activities it has helped to integrate the agricultural research efforts of the Third World. Increasingly it is also involving scientists from the developed countries in the agricultural problems of the developing ones, playing the part of marriage-broker. In building its own system, it has catalysed the formation of an emerging global one[208] from which the agricultures of both developed and developing countries have gained a great deal.

The strong crop focus of the early centres, with its emphasis on improvement through plant breeding, and on the broad adaptability of varieties – which meshed so well with the concept of an international centre – is now less in evidence in the system as a whole. The tidal enthusiasms of donor agencies for low input farming, scale-negative improvements, adverse environments, sustainability, etc. have taken their toll. Nevertheless, the comprehensiveness of its research and documentation services, the profusion of its networks, alumni and contacts in developing countries and the strength of the logistic support it can provide make the CGIAR system magnetic for agricultural scientists throughout the world. Its centres are increasingly the catalysts bringing them together to work more effectively on what is truly a world problem.

CHAPTER 9

# The fifth billion (1975–1986)

## 9.1 Introduction: resources and resourcefulness

Although the fifth billion was added to the world population in only 11 years, it began as a period of less apparent concern about the world food supply. There were, of course, continuing pessimistic prognostications but, for the first time since the FAO gathered its statistics, the food supply per head in developing countries as a whole showed a sustained increase (14%) in spite of a 26% increase in their population within 11 years. Reassuring also was the fact that this substantial increase in food production by the world as a whole did not require any increase in the area of arable land. There were small increases in South America (15%), Africa (4%) and Asia (0.5%) but these were offset by reductions in the more developed countries.

The average world yields of the staple cereals continued to rise over the 11-year period (Figure 17, p. 91), rice by 32% and wheat by 51%. These increases reflected the continuing impact of the dwarf cereals combined with a marked increase in the use of fertilizers in developing countries. The proportion of wheat and rice crops which were sown to dwarf varieties in developing countries more or less doubled over the period, while their use of fertilizer nitrogen, phosphorus and potassium increased by 440%, 317% and 210% respectively, offsetting the decline in arable land per head as populations increased. The area under irrigation in developing countries increased by 82%, one consequence being that development debt repayments became increasingly burdensome for them.

In the developed countries, the population increased by only 8.9%

and food production had no problem in keeping up. In fact, food surpluses were the problem in many developed countries, leading to reductions in publicly-funded agricultural research at the very time when, for the world as a whole, research was needed into ways of raising crop yield potentials after the full impact of the dwarfing genes is absorbed. Agricultural research had succeeded too well for its own health and, by and large, many of the apprehensions of the social science critics of the Green Revolution had not come to pass. Nevertheless, some of the Faustian bargains of high input agriculture became apparent during the accumulation of our fifth billion, and more biological approaches to the problems of agriculture, such as integrated pest management and the genetic engineering of plants, were emphasized.

The oil price crisis of 1972/3 promoted a phase of intensive energy analysis of agriculture. This brought home the realization not only of how far we have travelled from the self-sufficient farmer but also of the extent to which modern agriculture is dependent on energy from fossil fuel. For the major crops, however, there is still a return in yield of 3–4 joules for each additional joule of input energy. But perhaps the most important lesson to be drawn from the energy analysis of agriculture derives from our being able to put all the inputs and outputs on a common scale, of energy. When this is done we find that, at least for rice in Asia, maize in the USA and wheat in France, diminishing returns to input energy under even the most intensive systems have so far been avoided. Diminishing returns may apply to some forms of agriculture, as Turgot, Adam Smith and Malthus had proposed, but they need not characterize agriculture in general.

Nevertheless, the energy analysis of agriculture gave a nudge to the development of minimum tillage just at the time when suitable herbicides, such as glyphosate, became available. Apart from the savings in energy and manpower, and the greater flexibility and timeliness of minimum tillage, it can also permit the safe cultivation of steeper slopes, thereby increasing the potentially arable area. And by making faster crop turn-around possible, it can also increase cropping intensity.

Besides the land, water and energy resources of agriculture, the genetic resources of our crops also came under scrutiny during this period, particularly after the UN Conference on the Human Environment in Stockholm in 1972. Steps to stem genetic erosion led not only to their effective conservation, but also to a far wider concern for the preservation of global biodiversity, and the recognition of our wider evolutionary responsibility.

The development of the techniques of genetic engineering within the period under discussion made the genes of the whole biosphere, from virus to mammal, potentially available for crop improvement. Indeed, by the time our fifth billion had arrived, viral genes were beginning to confer virus resistance on potatoes, bacterial genes were enhancing insect resistance in cotton, and functional mammalian antibodies had been produced by tobacco. Molecular agriculture had become a field ripe for cultivation, not least by lawyers!

Along with these advances there inevitably came new hazards. Earlier generations have often cursed their forebears for the disasters resulting from their lack of foresight. Our successors may well curse us likewise, but they should at least be aware of the many and diverse efforts of agriculturists and others to foresee the adverse future effects of the new agricultural technologies[7,74,207]. However, as Winston Churchill once remarked: 'It is always wise to look ahead, but difficult to look further than you can see.'

## 9.2 Energy use in agriculture

When the downward trend in the real world prices for wheat and rice was sharply reversed by the oil supply crisis in 1973 (Figure 24, p. 135), the growing dependence of agriculture on off-farm sources of energy was suddenly highlighted. Energy analysis of agriculture became a growth industry.

The food sector generally accounts for 10–15% of total energy now used in industrialized countries, with only one fifth to one third of that proportion actually being used on-farm. In the USA, for example, about 17% of total energy use is for the food system, 6% for production and approximately equal amounts for packaging and for distribution. The consumption of energy on-farm has grown rapidly along with the rises in yield as mechanization was extended and diversified, as the dependence on fertilizers, pesticides and herbicides increased and as grain drying and crop irrigation by pumps and sprays became more common. As Howard Odum put it: 'man no longer eats potatoes made from solar energy; now he eats potatoes partly made of oil'.

The classical analysis of changing energy use in agriculture was made by David Pimentel and his colleagues at Cornell[157,158] on maize production in the USA, initially for six years between 1945 and 1970, but eventually extended to 1985. As is usual in such studies, the input of energy from the sun for crop growth was not included as a component of on-farm energy use because it would dwarf all the others, the intercepted solar energy being more than 99% of the total, considerably higher than Pimentel estimated it to be. On the output side the energy analyses stop at the farm gate, but they go upstream enough to include the energy 'embodied' in the manufacture and transport of machinery, fertilizers, etc.

Another convention of energy analysis concerns the farm workers: only the metabolic energy expended on their work is included, 5–8 megajoules (MJ) per day. For modern maize production this is a negligible item in the energy budget, less than 0.1%, but if the full charge for the farm worker's lifestyle support energy of up to 600 MJ per day were made, it would become a significant component.

Rather different conclusions have been drawn from Pimentel's studies depending on how the results are presented. The authors themselves compared the ratio of energy output/input against energy input and emphasized the implications of its fall with the intensification of agriculture. However, if we simply plot the output energy against the input energy we find that, overall, the maize crops have continued to return at least 3 more joules in additional grain for each additional joule invested in the crop, and substantially more than that from 1975 to 1985[58]. Only between 1964 and 1970 did there appear to be a sign of diminishing returns. But when Vaclav Smil and his colleagues[195] reanalysed the Pimentel data and allowed for the improvements in fertilizer manufacture and other changes, this apparent decline in efficiency disappeared.

Clearly, improvements in efficiency off-farm have helped to maintain the efficiency of energy use for maize production, but so also have improvements on-farm. For example, despite the four-fold increase in yield between 1945 and 1985, the use of liquid fuels per hectare decreased by 11% thanks to improvements in cultivation and weed control. The most pronounced change in energy-input terms has been for nitrogenous fertilizers, which rose from 6.7% of the total in 1945 to 31% in 1985. The second largest energy-using component in 1985 was irrigation (22%), up from 5% in 1945. Other marked increases have been for herbicides and grain drying.

Is US maize growing atypical? No other data set matches that of Pimentel and his colleagues. Instead of comparisons across time in one region, Flinn and Duff[65] compared the energy budgets of rice crops grown under twenty-one different conditions, mostly in Asia but ranging from traditional and low input to extremely intensive, both organic and conventional. Their results indicate the absence of diminishing returns to increasing energy input. Likewise in France, Sylvie Bonny[19] has found no decline in the efficiency of energy use for wheat growing in the Paris basin between 1955–60 and 1990.

In his book *Transforming Traditional Agriculture*, the distinguished economist T.W. Schultz[186] challenged the 'belief in a historical law of diminishing returns that holds uniquely for agriculture'. Of course, there are innumerable examples of diminishing returns to any one factor, e.g. nitrogenous fertilizer. But what energy analysis of agriculture permits is the reduction of the great variety of inputs and outputs in modern agriculture to a common base, energy consumed and energy produced. And what this has shown us, unambiguously, is that at least for rice, wheat and maize, the three main staple crops of the world, yields per hectare can be increased many-fold without diminishing returns to non-solar energy consumption setting in.

Did the oil embargo of the 1970s make a difference? Bonny shows that the trend in energy intensity of French agriculture as a whole reversed direction in the mid–1970s. Likewise, after a great leap in energy use for maize growing in the USA before the oil embargo there was a pause until 1975, since when the returns to input energy have been even greater than before. Modern agriculture is not prodigal of input energy, and next time you eat a slice of bread, please remember that much less energy was used in growing the wheat for it than in processing and distributing it for your convenience.

## 9.3 Minimum tillage systems

The scientific evidence in favour of reducing tillage, and systems for achieving that, had been around for thirty years before the oil embargo crisis of the early 1970s pushed agriculture more decisively towards minimum tillage.

Before mechanization, the horses and manpower required for tillage were at hand on the farm, so extra cultivations could have little or no cost. The guiding adage was: 'When the crop stands still, stir the soil.' Weeds were controlled and soil moisture was thought to be conserved. With tractors, however, each cultivation had an additional cost. Moreover, research at Rothamsted by Bernard Keen in the 1930s queried the efficacy of 'dust mulches' for water conservation and showed that ploughing was unnecessary if weeds could be controlled in some other way.

Indeed ploughing, that hallmark of good farming for more than a millennium, could cause substantial soil erosion, as recognized in the Great Plains of the USA during the 1930s when stubble mulching was found to reduce both wind and water erosion. When in 1943 an Ohio farmer, E.H. Faulkner, published *Plowman's Folly*, in which he suggested the abandonment of ploughing in favour of discing, there was outrage and controversy but the movement towards minimum cultivation had begun. It had two strands: economy of labour and fuel, and the conservation of soil and water. These are reflected in the range of its names, including zero-till, conservation tillage, crop residue management, low energy systems of cultivation and the ecological approach to soil management.

However, the key to the further development of minimum tillage was the introduction of herbicides effective enough to replace repeated cultivation for weed control. In fact, experiments by Keen and E.W. Russell in the 1930s on the effects of ploughing were made difficult by the lack of effective herbicides and led to a better appreciation of just how much crop yields were being reduced by weed competition. The first plant growth regulating herbicides became available in the late 1940s but it was not until 1959 that the first soil-acting, residual herbicide, atrazine, was introduced. This was followed in 1961 by the contact herbicide paraquat which opened the way to direct drilling. Glyphosate, discovered in 1971 and released in 1974, was the next major step along the herbicide trail, followed by several post-sowing, soil-active herbicides such as the sulfonyl ureas and nicol sulfuron, effective on perennial weeds in applications of only a few grams per hectare. Along with herbicide improvement by the agrichemical companies there has been continuous development of farm machinery for minimum tillage, much of it by farmers themselves.

The widespread adoption of minimum tillage techniques since the 1970s has depended on these advances and on clarification of their effects on crop yield. Under conditions of adequate rainfall, soil water and drainage, crop yield has been shown not to be reduced by the reduction in tillage. Where rainfall and the opportunities for irrigation are limited, yields are often higher with reduced tillage, associated with better water penetration and conservation. On the other hand, yields may be lower in poorly drained situations or where there are problems with pest, disease or weed control.

In the American mid-west the effects on yield of reduced tillage depend on the extent of soil coverage by mulch, high coverage raising yields above those of conventional tillage, low coverage reducing them. In the wetter conditions of north-west Europe, on the other hand, yields may be increased by removing all crop residues, e.g. by burning. In some environments crop yields have been found to improve as the number of years under reduced tillage increases. Rattan Lal[110] has grown two crops of maize each year for 17 years in West Africa, with higher yields from minimum tillage than on ploughed land, but with some problems of compaction which may require ploughing every 5–10 years. The enhancement of yields with minimum tillage can be obtained along with significant reductions in fuel and labour costs. With maize crops in the USA, the saving in total energy use for tillage and weed control is commonly 20 to 40%[168].

The reduction in soil erosion by minimum tillage can be striking, varying from 20- to 1000-fold across a range of environments[155]. An important consequence of this effect is that soils of much greater slope can be safely cropped with minimum tillage provided the mulch can be conserved. On a world scale, the ability to cultivate steeper slopes safely implies a considerable increase in the area of potentially cultivable land. But an equally valuable effect of minimum tillage, particularly at low latitudes, is the much faster turn-around between crops, thereby increasing the scope for multiple cropping. Another advantage of minimum tillage is that of more timely sowing, because the whole operation can be carried out over a wider range of soil moisture contents and not held up awaiting suitable conditions for ploughing. Moreover, the greater speed of minimum tillage means that about four times more land can be prepared and sown by each cultivator.

There are, of course, some disadvantages of minimum tillage. Some weeds are more difficult to control with it than with conventional tillage, as are some pests and diseases. Wetness may be aggravated in poorly drained soils, and there may be a slower warming up of soils in the spring, which is a problem in the Corn Belt[15]. On the other hand, the lower soil temperatures with mulching can be an advantage in the tropics.

Such differential effects have led to varied opinions of minimum tillage. It has been called 'a thing of the past, not a wave of the future', but also 'the greatest conservation practice of the twentieth century'. Although minimum tillage was the outcome of the serendipitous discovery of growth-regulating herbicides, its development has combined the contributions of scientists, conservationists, farmers and the agrichemical industry to transform the central act of agriculture, the cultivation of the soil.

## 9.4 Genetic resources

In agriculture no less than in other activities, success with old problems frequently begets new ones. We have already seen several examples of this, such as the environmental problems arising from the excessive use of DDT as an insecticide. Ironically, the rapid and widespread adoption in the late 1960s by farmers in the developing world of the dwarf wheat and rice varieties soon led to concern that the genetic diversity embodied in the old land races and wild relatives of these crops would be lost. Such genetic erosion was likely to be less readily repaired than soil erosion.

Back in 1923 Nicolai Vavilov had warned of the need to conserve genetic variation in crop plants in the face of agricultural change. Some years later Harry Harlan had repeated the warning, but it was the rapid spread of the dwarf cereals that brought the requisite sense of urgency to the issue. In countries where wheat and rice had evolved a great diversity of land races (i.e. traditional local varieties) many farmers were no longer growing them, but only the new higher yielding varieties more responsive to fertilizers. A single new rice variety, IR36, eventually occupied 10% of the world's total rice area, while selections from one wheat cross at CIMMYT were sown on a seventh of the developing world's wheat area at one stage.

In 1966 Otto Frankel, a geneticist and wheat breeder from Australia, brought together the International Biological Program's initiative on 'Plant gene pools', with its focus on the wild relatives of crop plants, and the FAO's initiative on 'Plant exploration and conservation', with its particular concern for the endangered land races. Conferences to clarify the priorities for conservation were held, and a panel of experts considered such matters as strategies for collection, evaluation and documentation, the recent improvements in long-term, low-temperature seed and tissue storage techniques, and the need for a global network for the conservation of plant genetic resources[67].

Funds for this work were not forthcoming, however, until the United Nations Conference on the Human Environment was held in Stockholm in 1972. Frankel addressed the conference on the dangers of genetic erosion, resolutions were adopted and the conservation of genetic resources became one of the world's enthusiasms. Funding and founding followed, with the CGIAR establishing its International Board for Plant Genetic Resources (IBPGR, now IPGRI) in 1973. As so often happens in human affairs, however, the wave of enthusiasm for the conservation of genetic resources began to be ridden by those with a variety of other agenda, which need not concern us here beyond the comment that among the issues raised were those of 'sovereignty of seeds' and of 'genetic debts', leading to FAO's International Undertaking on Plant Genetic Resources and, in 1993, to the United Nations Convention on Biological Diversity.

By now, however, the world's seed banks contain comprehensive collections of both the land races and the wild relatives of most major crops and of many minor ones. Across the various wheat collections, for example, there are more than 400 K accessions, probably representing about 125 K distinct samples from an estimated 90% of the world's land races plus about 60% of the variation among wild relatives. There are also more than 100 K accessions of rice, maize, barley, sorghum, beans and soybean[160]. The IRRI germplasm collection alone has more than 80 K accessions of wild and cultivated rices, with room in its base collection for 50% more. Of course, not all these samples will still be viable because, without frequent testing, seed banks can become seed morgues. But now that the most urgent collecting has been done, there can be more emphasis on the renewal and characterization of samples.

To date, land races have been used much more extensively than wild relatives in breeding programmes, in part because plant breeders believe they regain yield potential more rapidly after introducing a land race into their crossing programme than after using a wild relative. In rice, for example, resistances to most of the fungal (blast, sheath blight and brown spot), bacterial (blight and streak) and viral (tungro) diseases and to insect pests (stem borers, brown plant hopper, white backed hopper, green leaf hopper and gall midge) have come from land races. On the other hand, resistance to grassy stunt virus was initially found in only one out of almost 7 K accessions tested, and that was in a wild relative, *Oryza nivara*.

Although plant breeders tend to use wild relatives in a crossing programme only as a last resort, many valuable characteristics have been introduced into modern varieties from them, including resistances to pests and diseases, adaptations, quality factors and characteristics influencing reproduction. For several crops, as Jack Harlan[84] put it, 'all the cards in the deck are wild'. Moreover, with the help of tissue culture and embryo rescue techniques, wide crosses with less closely related wild species are now possible. As genetic engineering techniques develop, the ability to transfer single genes from quite unrelated wild species, indeed from any organism, without loss of adaptedness to modern agriculture, will also enhance the value of the wild relatives.

From the conservation of genetic resources for future agriculture to the conservation of biodiversity is in many ways a small step for mankind, although the modes of conservation may be quite different, more *in situ* and requiring more ecological insight for biodiversity. But, as Otto Frankel wrote in 1970: 'We have acquired evolutionary responsibility.'[66]

## 9.5 Integrated pest management: changing the paradigm

The introduction of the *Vedalia* beetle into California in 1889 to control the cottony-cushion scale of citrus was a spectacular example of the biological control of insect pests. At least, it was so until 1946 when the widespread use of DDT to control other pests killed off the beetles and led to a resurgence of catastrophic outbreaks of scale. When the use of DDT decreased, control by the *Vedalia* beetle was re-established, illustrating just how important predators and parasites can be in the regulation of insect pests[50].

Excessive use of insecticides like DDT, which destroy the natural populations of predators and parasites within sprayed crops, may also lead to outbreaks of new pests. Well before the release in 1966 of the dwarf rice variety IR8, a large proportion of the Philippine irrigated rice crop was being sprayed with insecticides provided free of charge as part of bilateral aid programmes. Following the release of IR8 and the intensification of rice farming, Filipino farmers considered it necessary and progressive to spray regularly with insecticides. Natural enemies within the crop were reduced and the brown plant hopper emerged as a serious new pest of rice in south-east Asia until largely curtailed by the release of the resistant variety IR36 in 1976.

However, insect pests have a habit of eventually overcoming their host's genes for resistance, and new sources of resistance have to be located and bred into new varieties. Several cycles of such breeding have been needed in rice for resistance to the brown plant hopper, just as they have been in sorghum for resistance to the greenbug and in wheat for resistance to the hessian fly[74]. Can we escape this treadmill of the evolution of pest resistance keeping up with the development of new pesticides and the breeding of varieties with new resistances? Probably never, but by using integrated pest management we should at least be able to prolong the useful lives of both new insecticides and new genes for pest resistance.

In 1959, Vernon Stern and Robert van den Bosch of Riverside and Ray Smith and Kenneth Hagen of Berkeley, California, combined to write a seminal paper on *The integrated control concept*[202]. In this they argued that, on the kind of evidence presented above, the controlling influence of natural predators and parasites should be preserved and engaged by minimizing the use of pesticides. This could be achieved only by discarding the prevailing 'clean crop' approach to agronomic hygiene, maintained by regular spraying with insecticides, and by tolerating the presence of pests up to a defined level at which *economic* injury to the crop actually occurs. Although it was this tolerance of pest populations up to a defined level, thereby enhancing the impact of natural enemies, that was the central innovation of this new paradigm[152], it also emphasized the integration of all synergistic agronomic and plant breeding practices. As it evolved, more emphasis was given to 'managing' the injury from insect pests and on optimizing long-term benefits and the economy of crop protection.

Although the 1959 paper put forward the concepts and set the agenda for integrated pest management (ipm), more than a decade of research was required to develop the requisite data and understanding for its components, especially the economic injury thresholds for the major pests of various crops. These are knowledge-intensive. Much research is needed to establish them, and considerable understanding and management skills to apply them.

Consequently, although it is widely agreed that only ipm provides the scientific and practical basis for a sustainable long-term control of pest problems, we still lack usable programmes for many important crop pests, even in developed countries[68]. The move towards the integration of pest, disease and weed control programmes will make for even greater managerial complexity. On the other hand, the need to ensure a long useful life for valuable resistance genes, such as that from *Bacillus thuringiensis* in cotton and other crops, should propel the wider use of ipm.

In developing countries the more limited research and educational base might retard the implementation of ipm, which could be a difficult challenge for small farmers with little education. However, the FAO 'Inter-country Program for integrated pest control in rice in south and south-east Asia', begun in 1986 under the leadership of Peter Kenmore, has shown that the small farmers of Indonesia can be effectively taught how to look at the rice paddy as an ecosystem and to manage it with smaller inputs of pesticides and fertilizer and with less variation in yield. With the Indonesian programme as an example, rice farmers throughout south and south-east Asia are now moving to ipm and putting world rice production on a more sustainable basis.

Kenmore insists that 'ipm is not a technology; it is a problem-solving process' through which 'National ipm programs replace investment in chemicals and their associated pest-surveillance systems by investment in people.'[109]

## 9.6 The genetic engineering of plants

As our fifth billion accumulated on earth, the genetic engineering of plants by means of recombinant DNA technology came of age and one of its founding fathers, Jeff Schell, suggested that it may get us around 'the constraint of time'.

In 1974 a gene from one bacterial species was cloned and expressed in a different species for the first time. The transfer of functional foreign genes, conferring antibiotic resistance from bacteria on transformed tobacco plants, was first reported in 1984. The crown gall agent (*Agrobacterium tumefaciens*) was used to transfer the resistance gene from bacterium to plant, and was the agent used in most early transformations. By 1987, however, DNA uptake by isolated protoplasts of both soybean and rapeseed cells had been successful, and by 1990 the 'DNA gun' had transformed soybean, cotton and corn[48].

Although the first transfers involved only easily-recognized marker genes, such as antibiotic resistance, potentially useful genes soon accompanied them. Resistance to injury by a variety of herbicides had been conferred on several crop plants by 1987, using various strategies such as detoxification of the herbicide or over-production of its target enzyme. For agrichemical companies, the prospect of being able to breed varieties of crop plants tolerant or resistant to the herbicides they produce was obviously attractive, and many such resistances have now been engineered.

The transfer to crops of resistance to insect pests was another early target, first achieved in 1987 by the transformation of tomato, tobacco and cotton for production of the potent insecticidal protein from *Bacillus thuringiensis*, commonly known as B.t. toxin. Engineering resistance to fungal diseases has followed similar lines, e.g. by incorporating genes for the enzyme which breaks down the chitin in fungal cell walls. Virus resistance had also been conferred on several crops by 1987, following a variety of strategies, e.g. by expressing the coat protein of the virus in the crop, using the genome of the virus itself, as well as that of the host, in fashioning resistance.

Although many further advances have been made, enough has been said to illustrate the versatility, the power and the speed of attack on several traditional objectives of plant breeding, particularly those of more immediate interest to agrichemical companies. What we have been considering, however, is simply the transfer to crop plants of alien genes with desirable characteristics. Traditional plant breeding procedures are then required to produce adapted and productive varieties, which in the case of the B.t. toxin in cotton has required almost 10 years from transformation to commercial use. Moreover, not all crops have been as susceptible to genetic engineering procedures as those of the potato family such as tobacco and tomato. Until recently transformations by means of the crown gall organism could not be used on the all-important cereals, and the need to rely on isolated protoplasts and the DNA gun limited progress.

However, genetic engineering offers the enormous advantage of being able to incorporate genes from any corner of the biosphere into crop plants. It also brings greater speed and specificity to plant breeding. Molecular techniques have shown that even after 20 generations of traditional backcross breeding in tomatoes, some unwanted DNA was still linked to the target gene. Moreover, DNA markers are greatly enhancing the efficiency of traditional selection for characteristics determined by many genes by allowing them to be followed through selection, and by indicating how many genes are involved, where they are located and how great are their relative effects[203].

Genetic engineering also brings risks. Just as excessive reliance on the new insecticides and herbicides soon led to the development by pests and weeds of resistance to them, with the loss of their effectiveness, so may the genetic engineering of varieties resistant to pests and herbicides lead all too soon to natural selection among the pests to overcome the genetic resistance of the crop, and to the acquisition of herbicide tolerance by weeds, particularly those related to the crop. More than 700 pests, 200 pathogens and 30 weeds have already developed resistance to agrichemicals[207]. Many of the genetic resistances bred into crops have already been bypassed by their insect pests, and there is no reason to suppose that genetically engineered resistances will be more permanent. Were several different resistance mechanisms to be combined, as is feasible, given time, their effective life should be greatly prolonged. Unfortunately, however, commercial imperatives may shorten the useful life of genes such as the B.t. toxin. Moreover, some of the engineered resistances may themselves impose a cost on crop productivity.

Apart from reducing yield loss due to pests and weeds, the impact of genetic engineering on the vital attribute of yield potential remains unclear. The yields of particular components, such as new starches, fatty acids, sterols, sweeteners such as thaumatin, polymers, antibodies and other pharmaceutical compounds could be enhanced many-fold. Nutritional value can be improved, as can storage and ripening characteristics. Rubisco or the light-harvesting proteins might be re-designed, as might the enzymes controlling sugar–starch partitioning, but whether the yield potential can be raised in such ways remains to be seen. The availability of molecular markers will certainly aid selection for yield potential, which is determined by many genes, but the real promise of genetic engineering may be with specific features better suited to industrial appropriation. Molecular diversity has become biology's new commodity.

CHAPTER 10

# The sixth billion (1986–1998/9)

## 10.1 Introduction: the recognition of limits

In 1998/9, just 200 years after Malthus first published his *Essay on the Principle of Population*, the world population will reach six billion. That is six times more than the population he estimated in 1798, and probably beyond what he considered possible. Yet, in the 12 years it has taken for the sixth billion to arrive – mainly as a result of increases in China, India and Africa – the FAO indices of food production per head for the world as a whole have shown no decline.

However, neither has there been a sustained rise in the food index for the world as a whole, such as that accompanying the preceding billion, although there has been some (10–14%) improvement in Asia and South America. Some observers have suggested, therefore, that surpassing a world population of five billion in the late 1980s may represent a turning point in the ability of global agricultural production to keep pace with population growth. Falls in food production per head have occurred in Africa and in some of the food-exporting developed countries where, faced with surpluses, governments have introduced measures to curtail production and reduce price supports and subsidies. The real world price for both wheat and rice remained generally low (Figure 24, p. 135).

The annual growth rate of the irrigated area in Asia (two thirds of that in the world) fell from over 2.5% in the 1970s to 0.4% in the late 1980s, along with a massive reduction in lending and assistance for irrigation by the development banks. Contributing factors include the already large public and foreign debt loads of many developing countries, the falling availability of unexploited irrigation potential, and the rising chorus of resistance to irrigation on environmental and public health grounds. But the major causes have probably been the reinforcing effects of declining world cereal prices and rising per hectare costs of irrigation development.

Svendsen and Rosegrant[204] ask of irrigation development: 'Will the future be like the past?', and answer in the negative. They do not foresee further major expansions of irrigation, in which case our sixth billion may indeed mark a turning point in our dependence on rising yields and intensity of cropping to match population growth. Coinciding and interacting with the downturn in the extension of irrigation, there has also been a period of stasis in the global use of N, P and K fertilizers, particularly in Europe. Nor has the world's arable area increased since 1986, small increases in the developing world being offset by decreases in the developed regions.

The accumulation of our sixth billion has also seen a decline in the real value of public funding for agricultural research in many countries and also in the international agricultural research centres. Moreover, there has been a noticeable shift of funding away from production-oriented research to a greater emphasis on the long-term sustainability of agriculture. While this is an appropriate goal for countries that have passed through their demographic transition, for those that have not it may deflect research resources away from the essential goal of raising food production in step with the rise in population. Greater emphasis on the sustainability of agriculture was stimulated by the 1987 Brundtland report on *Our Common Future*[226], in which sustainability was described as ensuring that development meets the needs of the present without compromising the ability of future generations to meet their own needs. These latter, of course, are not easy to assess ahead of the many advances that will undoubtedly occur. While the emphasis is timely, sustainability is too often equated with low input agriculture, which would certainly not be sustainable in a world with a population that will grow to at least 10 billion.

Another trend emerging in agricultural research is the greater role of commercial research, particularly in fields where intellectual property rights can secure the returns on advances, as in biotechnology. Besides contributing to a weakening of public funding for agricultural research, this trend fosters research on problems suitable for industrial appropriation, not necessarily those most urgently in need of understanding or solution.

The advent of the first plants with genetically engineered resistances to insect pests, viral diseases and specific herbicides coincided with the world population reaching five billion. As the sixth billion arrives, the first commercial varieties with these resistances are being released, and it is too soon to estimate their likely impact on crop yields. So far there have been no transformations likely to raise yield potential in the staple cereals. The most profound of these would involve an improvement in photosynthetic efficiency. Rubisco, the central enzyme in $CO_2$ fixation, is the prime, but by no means the only, target for improvement, and any assessment of success must take into account the optimization of the overall photosynthetic process in relation to nitrogen supply, water use and other aspects of the crop environment.

On the agronomic side there continues to be wide-ranging innovation but more in the direction of more active or environmentally acceptable compounds than in terms of novel functions or procedures. Computer models and expert systems are being used increasingly to aid decision-making by the farmers, helped along by better forecasting of weather and markets and the more readily available positional and surveillance data needed for 'precision farming'.

The period under consideration here has been referred to as 'the lost decade for economic development.'[2] Certainly it has lacked the confidence and optimism, the public investment backing, and the synergistic innovations of some earlier steps in world population. The requisite increases in food production have come mostly from further increases in staple crop yields, but these have derived mainly from the continuing effects of advances made 20–30 years ago. They may soon be exhausted, and the use of other resources may be approaching their limit. As we pass the six billion mark it is still not clear how the ten billion will be fed, nor that famines have been banished, to which those in Ethiopia in 1991 and North Korea in 1997 bear witness.

## 10.2 Agronomic innovations

Although there may not be a feast for us all at the world's table, the cupboard is by no means as bare as many pessimists would have us believe.

Agricultural scientists have under test a great variety of innovations and new approaches to old problems. Some are of recent origin, others have their roots in research done long ago but still awaiting complementary developments in other areas. Rather than focusing on just one or two of these, I shall explore their variety.

Fertilizers were initially broadcast over the field, then drilled in with the seed, then placed in bands below or beside the seed or pelleted with it to improve early access to nutrients by the seedlings. Seeds have long been coated with fungicides to reduce seedling disease, or with rhizobial preparations to enhance the nodulation of legumes. To these are now being added coatings with macronutrients such as calcium or phosphorus, micronutrients such as molybdenum, hydrophilic substances to attract water, peroxides to provide oxygen, and antidotes to pre-emergence herbicides[187]. The advantages of an early and healthy beginning of growth often persist through the crop cycle, so seed coatings are likely to become increasingly complex.

Fertilizer formulations are likewise undergoing continual change, particularly in the directions of greater specificity for individual crops or environments, less likelihood of fixation or loss, slower release, etc. Nitrogenous fertilizers are by far the dominant component, about 60% of the world total, yet only 30–70% of the amount of N applied is commonly taken up by the crop in the first season. In wet environments or under irrigation, especially on light-textured soils in cool conditions which limit crop growth, much of the applied N leaches out as nitrate, with adverse effects on water quality and human health. One approach to reducing these losses has been the application of inhibitors of nitrification, such as nitrapyrin, which inhibit the conversion of ammonia to the more mobile and leachable nitrate[163]. Given the overwhelming predominance of urea among the N fertilizers used in developing countries, losses by nitrification are a major problem in search of still better solutions. With irrigated crops or flooded paddies, on the other hand, there is also the danger of substantial losses of N by ammonia volatilization unless the fertilizer is placed in the soil. Research into the conditions exacerbating such losses is leading to more effective agronomic management techniques.

The age-old problems of pests and diseases of crops, which we have already visited several times, have provoked a great variety of new approaches arising from basic research to complement the continuing improvement of pesticides and their delivery, and the genetic engineering of resistances. In spite of the high cost of developing and obtaining approval for new agricultural chemicals, which may be slowing down their commercial release, there has been a quite remarkable thousand-fold increase in the activity of all three classes of crop protection chemicals (i.e. insecticides, fungicides and herbicides) during the 20th century[58], and there is no reason to suppose that trend is exhausted. One of its great advantages is that it has permitted a shift to low and ultra-low volume spraying, thereby eliminating the need for large volumes of water. It has also encouraged the development of more efficient spraying systems and made pest control feasible in situations where it was previously impractical. Many of the more recent pesticides also have reduced environmental persistence.

The range of biological control agents continues to expand, e.g. in the use of fungal pathogens, protozoa and nematodes to control insect pests. Pesticides are now being supplemented by behaviour-modifying semiochemicals, such as pheromone sexual signals and deterrents to egg-laying. Many pheromones are being used, for example, to bait traps for the pest or to confuse males as to the whereabouts of the females, bringing frustration to the pest and joy to the entomologist.

Plant hormones and inhibitors of their synthesis or action are also being used to modify crop growth or reproduction to an increasing extent. For example, growth retardants which prevent the synthesis of the gibberellins, which cause elongation, are widely used to enhance dwarfing in cereals, even in varieties with dwarfing genes, when high rates of nitrogen fertilizer application are used. Both the gibberellins and their inhibitors, as well as the auxins and ethylene, are being used to enhance or inhibit growth, flowering, fruit and seed setting and development, latex flow, etc[81]. Another family of plant hormones, the cytokinins, is used to control fruit size and delay senescence while yet another, abscisic acid and its derivatives, which control water use by plants, can reduce injury when applied after unseasonal cold spells. Growth regulators are also used as 'safeners' to protect crop plants from herbicides, e.g. in the direct seeding of rice.

Another area of agronomic advance is that of 'precision farming', also known as farming by soil, farming by satellite, site-specific management, variable rate technology, computer-aided farming, etc. As indicated by the range of names, it combines many elements: response to the variations of soil within a field, which requires accurate location by global positioning systems, plus detailed mapping or real-time assessment of soil, pest or weed variations; equipment which allows the application of fertilizers or pesticides at variable rates; and decision support systems. Although trials to date do not consistently show increased profitability[154], they do lead to reduced and more efficient use of fertilizers and pesticides.

Even this limited catalogue of current developments should indicate that the agronomic cupboard is by no means bare of innovations. In some instances plant breeding is contributing to the traditional tasks of agronomy, e.g. in the production of herbicide resistant varieties. Indeed, such reinforcing combinations of agronomic and genetic approaches can prove to be extremely powerful, as we have seen with the fertilizer–dwarfing gene–herbicide synergism and with integrated pest management. And as inputs and interactions multiply, farm management expertise becomes an ever more crucial element in crop production.

## 10.3 The challenge of improving photosynthesis

Photosynthesis, the process by which water and atmospheric $CO_2$ are converted into sugars through the trapping of solar energy by plants, is often denigrated because of its supposed inefficiency in commonly converting only a small proportion of incoming solar energy into biomass. In fact, under certain conditions, the photosynthetic process rivals the most efficient photoelectric devices in energy conversion, but it has to cope with continually changing and often stressful conditions, as well as with the continual adaptation and renewal of the photosynthetic machinery. Despite having been able to modify crop plants in many ways, mankind has not yet succeeded in improving the efficiency of photosynthesis. It remains a complex and engaging challenge to our ingenuity.

When the maximum rates of $CO_2$ exchange per unit leaf area are compared among wild relatives, old land races and new varieties of crops, in nearly all cases no increase has been found. Indeed, with several crops there has been a decrease[58]. This apparent paradox is partly due to a trade-off between leaf size and photosynthetic rate, with smaller leaves often having higher rates per unit leaf area but less photosynthesis per leaf and less water loss, of advantage in drier conditions. In modern agriculture, on the other hand, the advantage often lies with larger leaves giving faster coverage of the ground and denser canopies, despite the lower photosynthetic rates.

Besides the leaf area/photosynthetic rate trade-off, there are many more subtle photosynthetic adaptations which may have been selected for unconsciously long before we came to recognize their significance. For example, barleys grown in dry areas have a lighter leaf colour associated with less chlorophyll b, giving better adaptation to dry and bright conditions. Similarly, selection of cotton and wheat for bright but irrigated conditions appears to have resulted in more open stomata and cooler crop canopies, particularly in the afternoon, at the expense of water use efficiency. There has therefore been some increase in photosynthetic rate, but only because such irrigated crops are in a position to trade greater water loss for carbon gain.

One solution to the problems posed by hot, dry conditions, known as the $C_4$ pathway of photosynthesis, evolved and spread widely about 5–10 M years ago. Crops with this pathway include maize, sorghum and sugar cane, while many of the worst weeds at lower latitudes are also $C_4$ grasses. In essence the pathway involves an anatomical rearrangement in the leaves which allows tubes of cells to fix $CO_2$ into $C_4$ acids which are then transported to inner tubes of cells where the $CO_2$ is released and then fixed by the usual $C_3$ pathway into sugars which are loaded directly into the veins for export. The advantage of this re-arrangement is that the inner tube operates far more efficiently at the high internal $CO_2$ concentration so that the trade-off between carbon gain and water loss is greatly improved.

Evolution has provided other variations on the photosynthetic process, such as the crassulacean acid metabolism plants which reduce water loss by pre-fixing most of their $CO_2$ at night. Selection of crops for higher yield in different environments is likely to have involved a great variety of adaptive fine tuning of photosynthesis. But it is significant that all the natural variations in the photosynthetic process have involved solutions to the limitations imposed by the central $CO_2$-fixing enzyme, rubisco. Admittedly the enzyme in $C_4$ plants has slightly different properties from that in $C_3$ plants, but rubisco has undergone extremely slow change in the course of evolution.

Rubisco – more properly called ribulose–1,5-bisphosphate carboxylase-oxygenase – evolved about 3.8 billion years ago. Since that time it has been exposed to extreme changes in atmospheric composition, from far higher to much lower $CO_2$ levels, and to a rise in oxygen level largely as a result of its own activity. It is the most abundant enzyme in the world, constituting 20–25% of leaf protein, and its efficiency has undoubtedly been honed by prolonged and intense natural selection.

Nevertheless, some view it as an anachronism, its properties still reflecting the high atmospheric $CO_2$ concentrations of long ago. To others it is a 'feeble and confused catalyst'. They see it as feeble because of its low turn-over number and the limitation it supposedly imposes on plant growth in spite of requiring almost a quarter of the nitrogen in leaves; confused because, as its full name indicates, it is not only a carboxylase but an oxygenase and, at least in $C_3$ plants, the latter function substantially reduces photosynthetic efficiency. Photorespiration, the process resulting from oxygenase activity, was initially viewed as wasteful, and became an early target for improvement, especially because it was minimal in $C_4$ pathway plants. However, further research has suggested that it is not only unavoidable but essential to the maintenance of photosynthetic competence in bright light.

In these days of crop modification by genetic engineering, however, the 'improvement' of rubisco constitutes an irresistible challenge and would, if successful, be a supreme achievement in the annals of biological endeavour, even though rising atmospheric $CO_2$ levels are reducing the limitation of crop growth by this crucial enzyme.

## 10.4 The dilemmas of irrigation

The central dilemma of irrigation is how to reconcile the declining share of total water use for irrigation with the increasing need for irrigation to enhance the effect of other inputs and raise yields and food production at the required rate. Currently only about 18% of the world's arable land is irrigated, that proportion being 25% in developing countries but only 10% in developed ones. Of the total irrigated area, 63% is in Asia and 70% of it in developing countries. In Egypt all the arable land is irrigated, in Pakistan 77%, in China 48% and in India 33%.

Irrigation is almost as old as agriculture itself, and some ancient systems such as the Nile and the qanats of the Near East survive, indicating that irrigation *per se* is not unsustainable. The extension of irrigation gained impetus in mid-19th century India, initially in the south and then in the Indus and Ganges watersheds. It then took off in the 1960s, reaching 2.7% per year in the 1970s since when it has declined to less than 0.4% per year.

Thus, the period of most rapid expansion of irrigation coincided with the spread of the dwarf varieties of wheat and rice and with the greater use of nitrogenous fertilizers (Figure 17, p. 91). Indeed, the extension of irrigation was a key component in the success of the dwarf varieties, by making heavier dressings of fertilizer worthwhile. Over 70% of the gains in wheat production from the use of dwarf varieties came from the irrigated areas[32], and nearly all of those for rice. Besides being crucial to increase in yield per crop, irrigation is also the key to further increases in cropping intensity, i.e. in the number of crops that can be grown per year, which may be an important source of greater food production at low latitudes in the coming years. There are many uncertainties in calculating the proportion of the world's food supply currently contributed by the irrigated areas, most estimates ranging between a third and a half.

However, water is an increasingly scarce and valuable resource, and the share of it used for irrigation has declined progressively from 90% in 1900 to less than 70% today. This decline reflects the increasing use of water by industry (now about 25%) and municipalities. In recent years there has also been a precipitous decline in irrigation investment in developing countries and in lending for it by the World Bank. In part this is due to the very success of irrigated agriculture in increasing the production of wheat and rice and in lowering their real price, with the consequence that returns from earlier irrigation schemes are falling as the cost of new ones rises, another dilemma of irrigation.

There is debate, therefore, as to whether the era of rapid expansion of irrigation has ended, with future emphasis to be placed on improvements in the efficiency of water use, or whether there will have to be further heavy investment in new irrigation systems as well as in the maintenance and improvement of existing ones. There is also active debate on whether returns are greater from investment in dryland as against irrigated agriculture, and from small rather than large irrigation projects. Private sector tubewell irrigation, so extensive and successful in south Asia, can enhance the efficiency of surface water irrigation both as a supplement at times of highest returns, and by rescuing percolation losses that recharge the aquifers[173].

We are in acute danger of biting the hand that feeds us on a world scale. Irrigation is currently getting a bad press, for several reasons. The limited life of many schemes offends the present emphasis on sustainability, as does the common problem of salination when drainage is inadequate, as it seems to be on about a quarter of all schemes. The fact that far more prime agricultural land is lost to urbanization than to salination is usually overlooked. The lowering of water tables by tubewell pumping is also a concern, but on the high plains of Texas the lowering caused by irrigation of sorghum has been replaced by that caused by urban demands, without objection.

Most irrigation schemes are inefficient in their use of water, i.e. in the proportion of the water released from the dam that is transpired by the crop. This is commonly said to be about 50%, ranging from 10–80%, and is clearly a prime target for improvement. Losses begin with those in the reticulation canals and end with those from inefficient or excessive applications. Irrigation can be prominent in the energy budgets of crops and in some cases substantially lowers their energy use efficiency. It may also pose health problems, such as malaria and bilharzia.

Can the central dilemma of irrigation be resolved simply by increasing the efficiency of water use in existing schemes, as well as maintaining them more effectively? There is certainly scope for improvement, but given the relatively low efficiency of even the well-designed and operated Columbia Basin project, the potential for improvement may be modest[174]. Reticulation losses could certainly be reduced, e.g. by lining or covering canals, but at considerable expense. In the irrigated fields there is also considerable scope, using the great variety of more sophisticated systems of application. These are widely used in developed countries but are mostly beyond the resources of the developing country farmers who use 70% of the world's irrigation water. The conjunctive use of surface water and ground water, though often recommended, is rarely practised.

Between the irrigation engineer and the farmer there has often been a gap in communication which has been the source of many past inefficiencies, because regular programmed releases of water from the dam do not match the varying requirements of crops as they develop and as weather varies. It is here that much progress has been made in recent years, as well as in techniques for assessing crop needs and of combining the distribution of fertilizers and of agrichemicals with improved methods of water application.

Will such improvements be sufficient to allow food production to keep pace with world population growth over the next few decades without a substantial expansion of irrigation capacity? I doubt it. There would have to be a renewed expansion of the arable area which could have as deleterious an environmental impact as the extension of irrigation. So we shall probably see more investment in irrigation, but within the context of providing a more sophisticated and responsive irrigation service, with more equitable and timely deliveries under specified rights and managed by smaller entities, with charges for deliveries[204].

## 10.5 Sustainability, the new watchword

Agriculture has always had elements of a Faustian bargain in its trade-offs between productivity and sustainability. Many generations of farmers have sought to improve rather than mine their land, as exemplified by the old saying: 'Live as if you would die tomorrow, farm as if you would live forever.' But as the environmental impact of modern agriculture has spread beyond the farm gate, so has awareness grown of the need to consider more global aspects of its consumption of resources and the long-term sustainability of our food production systems. As Holmberg *et al.*[101] put it, 'Sustainable agriculture is not a luxury … . When an agricultural resource base erodes past a certain point, the civilization it has supported collapses … There is no such thing as a post-agricultural society.'

Among the earliest concerns was how to maintain soil fertility, variously solved by the long forest fallows of shifting cultivators, the shorter arable fallows of early agriculturists, the continual influx of nutrients with irrigation water from the Nile or in wet rice culture, the rotation of legumes with the cereal staples, and the development of alternate cropping and grazing rotations, as in the Norfolk four-course. Several of these systems, such as agriculture along the Nile or wet rice culture at Tsukumino Moor in Japan[76], have proved their sustainability over several thousand years.

Soil erosion has also been a continuing concern, as portrayed by Plato more than two thousand years ago, and recognized as a public responsibility in the USA in 1935. Stubble mulching and the development of minimum tillage techniques have done much to reduce, but not eliminate, erosion in developed countries. However, it continues to be a serious problem in many developing ones where population growth and poverty force many to scratch a living from slopes too steep or land too poor to be cultivated in a sustainable way. Overgrazing and deforestation (which has increased in rate to about 1% per year since the world population reached four billion) are the cause of almost two-thirds of world-wide soil degradation in recent decades. Over-exploitation for fuel wood accounts for a further 7%, and poor agricultural practice for 28%. Salination and waterlogging, known since the time of the Sumerians, are still serious problems for irrigation schemes, causing the loss of about 1.5–2 M hectares annually. Most commonly this is associated with inadequate drainage, but the limited life span of some irrigation schemes in arid environments should be recognized at the outset.

New concerns for the sustainability of agriculture arose from the widespread use of fertilizers and agrichemicals, initially in the developed countries where applications were heaviest, but now in almost all. However, procedures for more efficient and moderate use have been developed and less polluting, less persistent and more effective pesticides are replacing earlier compounds. The mode and timing of their application have also been improved.

The intensification of agriculture is often viewed as the major source of anxieties about its sustainability. Without such intensification, however, the problem would be far worse because the increasing world population could then be sustained only by a massive increase in the area of arable land, the supply of which would run out long before ten billion of us arrived. Intensification *per se* need not cause environmental degradation but it certainly increases the rate at which mineral resources are used.

Although the high yielding varieties have contributed to sustainability in a number of ways, the spread of the dwarf cereals was seen as a threat to a different kind of resource, the genetic resources related to and available for the further improvement of crop plants. Timely action on that problem has resulted in our now having very extensive collections of the wild relatives and land races of our main crop plants in effective conservation.

Currently the emphasis is shifting from the sustainability of agriculture to that of the whole biosphere and of the global environment. Concern is now being expressed about the growing extent of the human appropriation of the products of terrestrial photosynthesis, approaching 40% according to one estimate[214], and of the world's fresh water[161].

Clearly, the framework for discussion of the sustainability of agriculture is evolving as earlier concerns diminish and new ones arise. Poverty and underdevelopment are continuing major sources of environmental degradation, for which the poor often have no realistic alternative. With development, on the other hand, come greater pressures on resources. Sustainable agriculture is not and cannot be equated with low input agriculture in a world with six to ten billion people to be fed. Neither simplistic nor short-term solutions will suffice and long-term experiments are needed to provide the data on which long-term solutions can be based, particularly for the lower latitudes, as has been shown with multiple rice cropping at IRRI. Research on sustainability requires sustainability of funding for such research, and it is notable that the endless calls for the one are not matched by the supply of the other.

# CHAPTER 11

# What the world eats now

## 11.1 Introduction: food for thought about food

In this chapter we look at several aspects of the present world food situation and begin to wonder where we go from here. But the present is often a confusing basis for extrapolation, with so many uncertain, even conflicting trends, some of which are considered below. Just as literature is searching for directions after post-modernism, so is agriculture assessing its post-Green Revolution futures from a complex and perturbing base.

Despite a doubling of world population since 1960, the food supply per head for the world has increased, calories by 13%, protein by 8%, and both by even greater margins in the developing countries as a whole. The *proportions* of malnourished people and of underweight children in them have fallen. Yet there remains an unacceptably high number of the absolutely poor and hungry, up to 800 million or so, in a world which is producing enough food for all but feeding almost half of its cereal grain production to animals.

As Amartya Sen showed in his thought-provoking book on famines[188], the problem of food insecurity is less one of insufficient production than of lack of 'entitlement', i.e. of the means to command food: 'Hunger and famine have to be seen as economic phenomena in the broadest sense ... and not just as reflections of problems of food production.'[189] In the Bengal famine of 1943, the Ethiopian famines of 1973 and 1982, and the Bangladesh famine of 1974 there was no decline in food output or availability per head, indeed in the Bangladesh famine these were higher than in the preceding and following years. But the food which could not be paid for locally went to other markets. On a world scale such lack of entitlement by the poor results in the demand for food being less than the need. Thus, while many of those with low incomes have benefited from the fall in cereal prices in recent years, the absolutely poor may not have. Food aid is equivalent to only about 1% of cereal production by developing countries, and doesn't solve the problem.

Hunger amid plenty is only one of the confusing elements of the present food scene. Another is the extent to which the flow of grain from the developed to the developing countries is likely to be a permanent feature of what Millman *et al.*[137] refer to as 'the emerging global food system'. They foresee a growing international interdependency in this, fostered by trade liberalization, improved transport capacity, information flow and infrastructure, more food processing and greater urbanization. However, the developed countries may well retreat from their highly subsidized surpluses of grain, and the urge for self-sufficiency in food and feed production may reassert itself in many developing countries.

Another uncertain element in the present scene involves the trade-offs between the intensification of agriculture and the conservation of the environment and its resources. In developed countries in recent years there has been much discussion of 'alternative agriculture' and some retreat, especially in Europe, to reduced inputs, e.g. of N fertilizers since 1989. A trend towards more sparing fertilizer use is also apparent in countries like Taiwan and Korea. In most developing countries, however, further intensification of agriculture is the most assured route to greater food production and the only real alternative to continuing deforestation and the extension of cultivation to less suitable and more vulnerable areas.

Another confusing element is the recent trend towards disinvestment in publicly-supported and publicly-available agricultural research in developed countries, associated in part with the accumulation of agricultural surpluses. Increasing commercial support for research in areas such as genetic engineering, where advances can be protected by intellectual property rights, may also be involved. However, there are also many areas of agricultural research not suited to industrial appropriation yet essential to the progress of agriculture.

Many economists appear to be relatively sanguine about the prospects, based on the extrapolation of recent trends, for further improving the food supply in developing countries, except possibly in sub-Saharan Africa. As a crop physiologist I am sceptical of such extrapolations, but the unflagging power of empirical selection for higher yields, and continuing agricultural innovation in the face of population and economic pressures, may well justify the economists' assumptions.

## 11.2 Food production and our global diet

An Iron Age man buried in a Danish bog lost some of his bones to the acid waters but, by way of compensation, his skin was tanned and superbly preserved, as were his stomach contents. So we know that his last meals included at least 60 plant species, many now considered weeds but also barley and linseed, and no meat.

Knowing what the world as a whole eats today is not so easy, but can be approached from two directions. The first, less direct one is based on the production statistics for crops and animal products published annually by FAO. These must be treated with some reservation, especially those from developing countries because under-development is also reflected in national capacities for data gathering and synthesis. At the farm level production may be under-declared for a variety of reasons, such as household use, stockpiling, black markets or the avoidance of taxes and requisitions. There are also problems with subsistence holdings, repeatedly harvested crops such as cowpeas, sporadically harvested ones like cassava, yams and bananas, and losses from storage. Even at the national level there may be reasons for under- or over-estimating production at times. Beyond those estimates, there are the further uncertainties of how much of the harvest is retained for seed or used for animal feed, brewing or other industrial purposes.

The second approach is the periodic surveying by FAO of what is actually eaten in the sampled households of various countries. This is fraught with the problems of all such surveys on a world-wide scale, and can be undertaken only occasionally, the most recent (fifth) survey having been published in 1987[62].

The first approach is illustrated in Figure 26, in which the overall length of each bar represents the average world production of the major food items in recent years (1993–95). The amounts of water and inedible matter (e.g. rice husks or sugar canes) in each item and the amounts

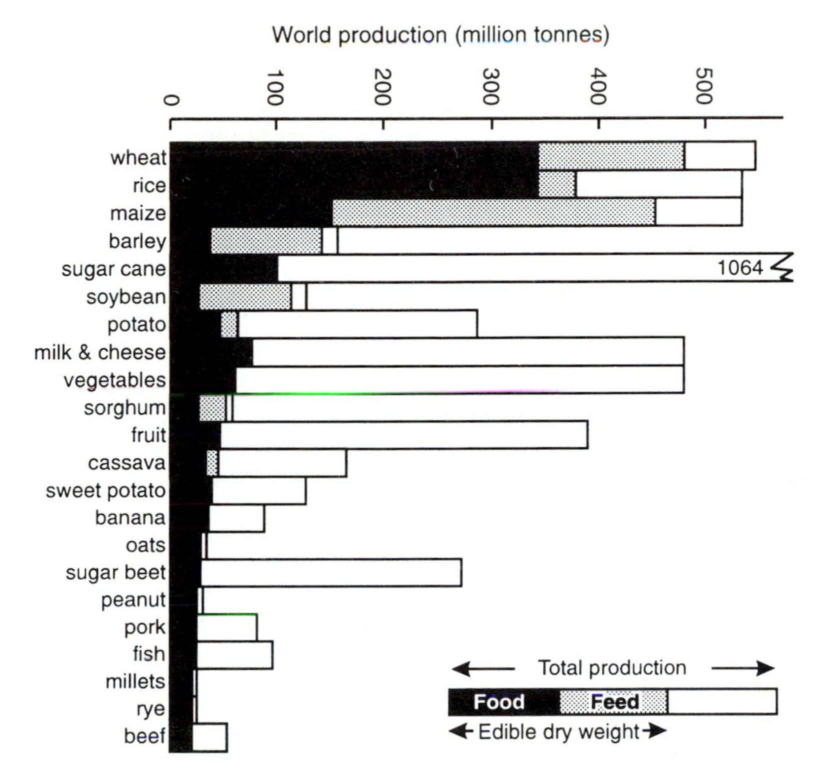


**Figure 26** World production of the major foods in 1993/95. The overall length of each bar represents the harvested weight of the commodities. The more lightly shaded areas represent the amount used for animal feed and the black bars the amount available for food in terms of estimated dry weight. (Data from FAO Production Yearbooks.)

used for animal feed and other purposes are then subtracted, to leave the amounts available for food, indicated by the black bars.

These estimates suggest that, for the world as a whole, 84% of the dry weight of the world's food derives directly from plants, closely comparable with the proportion of dietary calories estimated from the world food survey. Moreover, a large fraction of the remaining 16% from animal products is based on feed production by maize, soybean, wheat, barley, sorghum, cassava, sweet potato, cottonseed, etc. To a far greater extent than is generally recognized in developed countries, a remarkably small number of crop plants provides most of what the world eats.

Among these the cereal staples are pre-eminent. From the data for

Figure 26, I estimate that 54% of the world's food dry weight came directly from cereals, to be compared with the World Food Survey figure of 50.2% of diet calories. Rice and wheat alone contribute more than 44% of the dry weight, in about equal amounts. World maize production is comparable to that of rice and wheat, but a far higher proportion of it (about two thirds) is used for feed. Much of the barley is also used for feed.

Of the other plants, root crops contribute less than 10% of food dry weight, with potatoes the fifth most important source, behind sugar cane. The two sugar crops contribute 8.2% of edible dry weight (cf. 9.1% of diet calories). Vegetables and fruit contribute about 7.1%, the legume pulses (excluding soybean and groundnut) less than 3%. Animal products contribute about one sixth of dietary dry weight, 16.3% of the calories but a much valued 35% of the protein in the world's diet.

The two approaches to assessing the global food supply agree on the overwhelming role of crop plants in supplying us with energy, as well as with two thirds of our protein directly and most of the rest indirectly. With economic development there is usually a sharp rise in the proportion of animal products in the diet. Because this requires a more than equivalent increase in feed production, the rise is more apparent in the dietary surveys than in crop production returns, where it appears as a rise in the relative contribution of feed crops such as maize and soybean. In the most recent world food survey, 60% of dietary protein came from animal products in the developed countries compared with only 22% in developing ones, reflected for example in the rising proportion of maize used for feed as average national income rises. As the staple crops increasingly dominate agricultural output, grazed pastures become relatively less important, as do the oceans, which provide less than 1% of dietary calories. Only 19% of the world fish catch currently comes from aquaculture, but that proportion is likely to grow and much of it will also depend on feed crops.

Over the next 20 years or so, cereal consumption per head in the developing countries is projected to increase by over 2% per year, due to rising demand for animal foods rather than for greater direct consumption[103]. However, the production of cereals per head in the world, which rose from 277 kg per year in 1948–52 to 370 kg per year in 1976 has not increased since then, indeed it has decreased, and cereal production may be struggling to keep up with population growth for some time yet. Across the last three world food surveys, between 1961 and 1981, the main changes have been declines in the contributions by legume and root/tuber crops, matched by increases from vegetable oils, sugar and vegetables and fruits.

## 11.3 Regional variations in food supply

The proposition that 'the earth is one but the world is not'[226] is nowhere more apparent than in our food and its abundance. Although many crop plants were rapidly redistributed around the world after the voyages of Columbus, pronounced regional and national differences in diet remain as a result of differences in agricultural and culinary traditions, in soil and climate and in stage of economic development.

Although wheat and rice are equally important as staple foods for the world as a whole, wheat is by far the more significant in most developed countries whereas rice is twice as important as wheat for the developing countries and for Asia as a whole, as also for China and India. In Bangladesh rice production is 25 times greater than that of wheat, while Indonesian wheat production is still not sufficient to be listed in the FAO Production Yearbook. Asia still has the highest dietary dependence on cereals, which provided 67% of the *direct* intake of calories there in the Fifth World Food Survey. But even within countries there may be considerable regional variation in the staple foods. Although rice is the staple food for India as a whole, dietary surveys have shown that to be so for only half of the population: millets and sorghum are staples for one quarter of the people, and wheat for almost as many.

The developing region with the lowest dependence on cereals, less than 40% of dietary energy, is Latin America. This is the developing region with the highest proportion of digestible energy from animal products (17.1% cf. 5.8% in Asia and 31.7% in the developed countries) and from sugar (17.2% cf. 8.4% in Asia and 13% in developed countries), reflecting to a large extent the respective influences of Argentina and Brazil. Maize is the dominant staple for Latin America as a whole and especially in Brazil and Mexico, followed by wheat in Argentina, rice in Brazil and potatoes in the Andean countries.

In the Near East the wheats are the dominant staple, followed by barley, but in Africa the situation is more complex. North Africa is like the Near East, but for Africa as a whole maize and cassava are the most important staples, followed by rice and yams in the west, wheat in the east, and sorghum and millets in the drier regions.

When it comes to the average supplies of calories and protein per head per day there are large differences by continent, by country and, of course, by individual. For the world as a whole the energy available per head per day (now ~ 2700 kilocalories) increased about 18% from the early 1960s to the 1990s. Between countries there is a more than two-fold range from less than 1800 kilocalories in several Sub-Saharan countries to more than 3800 in Ireland, ironically the scene of a severe famine 150 years ago. The geography of dietary energy supplies at the time of the last world food survey is indicated in Figure 27, in which the relative size of countries is shown in proportion to their population in 1979–81. Since then, supplies of digestible energy per head have risen most rapidly in Asia (by 13.6%) reflecting an increase of 17% in China and 22% in India combined with falls in several smaller countries. In Latin America there has been little change overall, improvements in some countries balancing falls in others. Supplies of digestible energy per head have also held steady for the Near East and North Africa, but only because of their increasing dependence on imported food. The region of greatest concern is sub-Saharan Africa where food production has not kept pace with population growth, resulting in an almost 25% decline in average supplies per head in recent years, but being much more severe than that in many of the smaller, poorer countries.

The average consumption of protein for the world as a whole in 1992 was 70.8 g per head per day, 9% more than in 1970, with one third (35%) coming from animal sources. The regional averages range from 56.0 g (and falling) in Africa to 100.6 g (and rising) in Europe. Protein consumption per head is increasing in all regions except Africa, by 21% since 1970 in Asia. Also increasing is the proportion from animal sources, which ranges from 21% in Africa to 58% in North America, and has increased by 74% in Asia since 1970. China and India make an interesting contrast. With only one sixth of it coming from animal sources, India's daily protein consumption per head is only 58.1 g and has increased by only 10% since 1970, whereas China's has risen by 40%, with most of the increase from animal sources.

The nutritional fate of our fellow human beings varies by region, by country, by district, by year, by tradition, by gender and by income. It can be ameliorated to a small extent by aid and trade, but for most of our additional billions it will depend primarily on local production. And in that the earth may be one but the world is not. Malthus stalks through much of Africa while Boserup is at work in China.

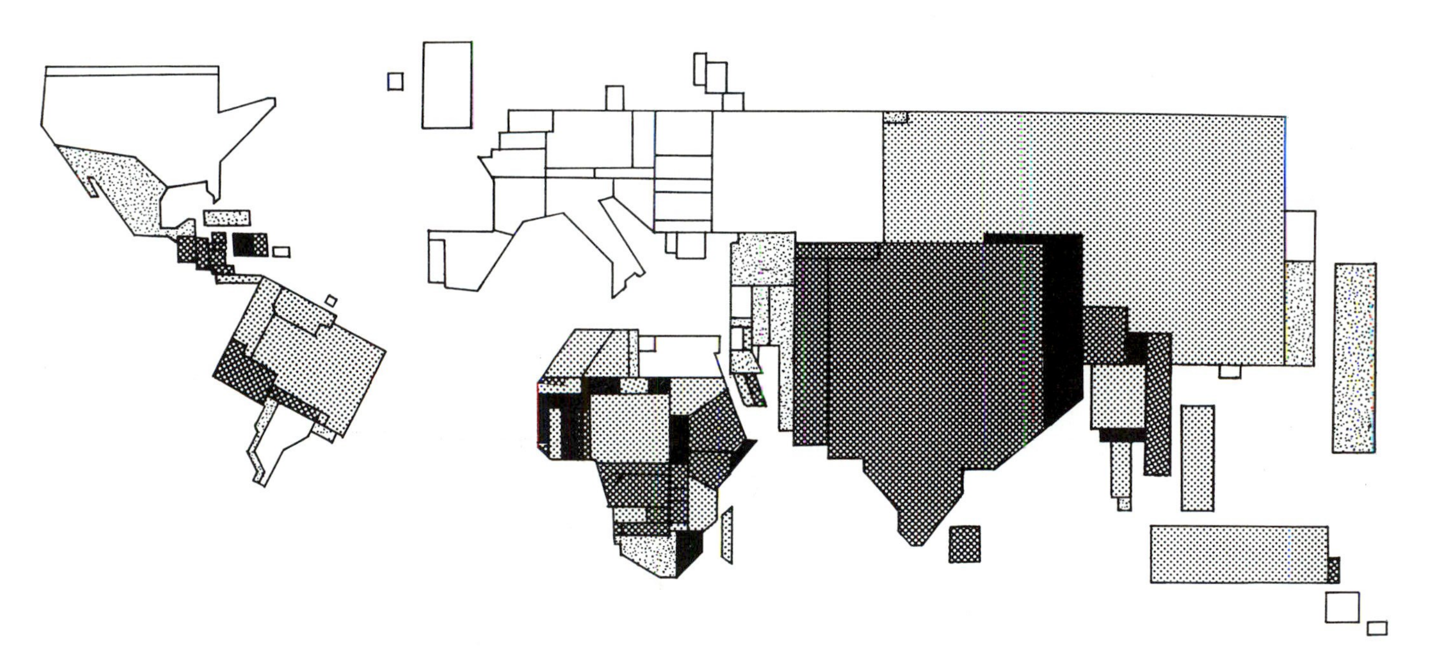

**Figure 27** The geography of dietary energy supplies per head, 1979/81. Countries are drawn in proportion to their population, while the hatching indicates average calorie intake from over 3000 (no hatching) to less than 2000 (solid black) kilocalories per head per day (Adapted from FAO[62].)

## 11.4 Hunger, malnutrition and poverty

Hunger and plenty co-exist today as they have throughout human history, but with less reason. The difference is that enough food for all is now produced in the world, even in times of local famine, yet the poorest of the poor, up to 800 M of them, still suffer chronic undernutrition. Despite the ringing Declaration of Human Rights in 1948 that 'everyone has the right to a standard of living adequate for the health and well-being, of himself and his family, including food'; despite the International Covenant of the United Nations in 1966 formalizing the right to food as a basic human right; and despite the closing declaration of the World Food Conference in 1974 that hunger would be eliminated within a decade, up to one fifth of the people in developing countries are still chronically undernourished.

Hunger and poverty are closely linked at the bottom end of the income distribution curve, both between countries and within them. Even in countries with a high average income, the poorest may go hungry, but to a lesser degree than in developing countries. Thus, a high level of national economic development may not eliminate hunger, but for the less developed countries of the world it could go a long way towards reducing the shocking incidence of undernutrition.

Estimates of the scale of the problem, and of progress on it, vary to some extent between the various surveys, depending on the criteria used. There is general agreement that the *proportion* of chronically undernourished people in the developing countries is falling, and the data in Figure 28 suggest that the absolute number is also, mainly due to the great advances made in China since the 1970s. It is still high in south Asia and, as a proportion of the population, highest in sub-Saharan Africa.

For the developing countries as a whole, the proportion of the population suffering undernutrition had fallen from 36% at the beginning of the 1970s to about 20% by 1990, and is projected by FAO to fall to 11% by AD 2010[2]. However, the proportion of malnourished children is much higher. This was estimated to be about 34% in 1990 – and almost 60% in south Asia – and is expected to be about 24% by AD 2020[175]. Progress in the 1970s reduced the absolute number of underweight preschool children, but that number has since increased.

Over half of the world's underweight children live in south Asia, where this chronic disjunction demands help. However, the slow

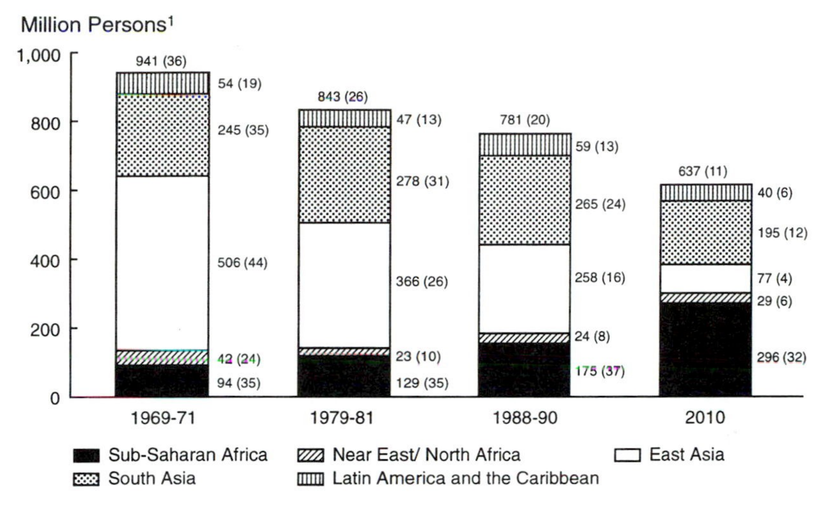


**Figure 28** Trends in chronic undernutrition in developing countries, by region (Alexandratos[2]).

progress in that region, evident in Figure 28, tends to be overshadowed by the escalating problem in sub-Saharan Africa, with its high birth rates, where about a quarter of the children are underweight and a third of the population is expected to remain chronically undernourished even by AD 2010[2].

Although protein–calorie deficiency is the most widespread form of malnutrition, others are also important. Anaemia due to iron deficiency affects 60% of women in south Asia, where current trends in dietary iron supplies are downwards. Overt cretinism from iodine deficiency affects six million people, with more than 200 million at risk of some degree of mental deficiency. Vitamin A deficiency in at least 40 countries causes eye injury to 14 M pre-school children, blinding half a million of them each year. Nutritional stunting and wasting, common among women in Asia, have persistent effects on later generations.

Besides the link with poverty, undernutrition also has regional and climatic dimensions, partly evident in Figure 27 showing the calories per head per day available in different countries. Undernutrition tends to be more prevalent at low than at high latitudes, and is much higher in the warm semi-arid tropics (49% of the population) than in the cool subtropics with winter rainfall (17%)[190]. Diets, as determined by region and climate, also have an influence. Although starchy roots, tubers and plantains contribute only 6% of the energy in developing country diets overall, that proportion rises to 40% in many of the poorest countries in Africa, to 13% in Latin America, and to far higher levels for the poorest, especially in bad years.

So far we have been focusing on food poverty, in which only the poorest households are in real want. Regional famine is a less predictable, more sporadic and dramatic nutritional stress. As such, famines attract the attention of the media and emergency aid, which chronic food insecurity does not. Yet although precipitated by drought, disease or war, it is poverty that makes famine possible, and it is the root causes of poverty that must be attacked if famine is to be prevented.

In *Our Common Future* the Brundtland Commission[226] wrote: 'We recognize that poverty, environmental degradation and population growth are inextricably related and that none of these fundamental problems can be successfully addressed in isolation.' But by focusing on the reduction of poverty and chronic undernutrition in the developing countries, which is within our grasp, the problems of population growth and environmental degradation could become more manageable.

## 11.5 Animal food and feed

Two common targets of concern about modern agriculture are its heavy dependence on inputs such as fertilizers, pesticides and irrigation on the one hand, and the great extent to which our staple crops are used for animal feed on the other. For the world as a whole, about 44% of cereal production is used to feed animals, three quarters of it in the developed countries[183,219]. In a world where 800 million people are chronically undernourished, this high proportion is a frequent cause of outrage. It also suggests that ten billion people could already be fed, given equitable distribution.

However, the consumption of meat in the developing countries is increasing by 5% per year, and that of milk by 4%. Brazil already uses more cereal for feed than for food, and in newly industrializing countries like Taiwan and the Korean Republic, rapid growth in incomes has been matched by rapid increases in the consumption of livestock products based on imported feed grains. The consumption of pig meat in

China increased three-fold, and of poultry four-fold, between 1980 and 1994, with the result that 20% of its grain production is now used for feed.

Besides reflecting dietary wants, if not needs, there are several advantages to increasing animal production in developing countries. For many small farmers livestock raising is an important source of income and employment. It is also an effective use of brans and oilseed meals. Bran accounts for one sixth of the cereals used as feed in the world, and for about 80% of the wheat and rice used for feed. Animal feeding is also a preferred way of disposing of crops surplus to the farmer's requirements.

Nevertheless, for the world as a whole, pasture and forage still provide about three quarters of the energy in livestock diets[219], compared with one sixth from grain and 8.5% from meals and by-products. Cattle, sheep and goat meat production is still largely pasture-based, whereas poultry, eggs and pork are largely feed-based, using about 60% of the feed grain. The escalating consumption of food of animal origin in developing countries means that the increase in crop production must exceed that in population, enhancing the need to raise yields. Sarma[183] estimates an average ratio of feed input to food output by weight of 2–3 for chicken meat, 4–6 for intensively fed pig meat and even higher for lot-fed beef.

The large regional and national differences in calorie, protein and fat supplies per head per day clearly relate to average income, as Figure 29 indicates. Among the developed countries, average income has relatively little effect on energy and protein intakes, New Zealand, France and Ireland having the highest consumption of animal protein but not the highest incomes, for example. But as we move to the left of the figure, towards lower incomes among the developing countries, animal protein intake falls rapidly to very low levels. The fat component of the diet tends to vary in parallel with the protein, ranging from over 140 g down to 47 g per day, with the proportion of animal origin falling from 60% in Europe to 24% in Africa.

For the world as a whole two thirds of all the maize and other coarse grains produced are used for feed, this proportion rising to about 95% for sorghum in the developed countries and in Latin America. By contrast, only about 30% of wheat and 8% of world rice production are used for feed, and about a quarter of the production of cassava, other roots and tubers and the pulse crops[1]. The use of cereal grains and

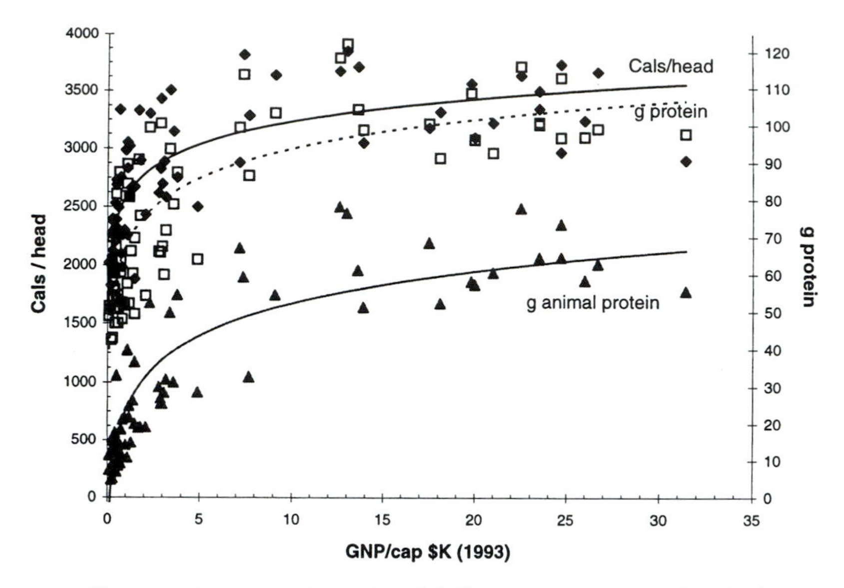


**Figure 29** Increases in national daily consumption per head of calories (◆), protein (□) and animal protein (▲) with increase in GNP per head. (Data from World Bank and FAO.)

by-products as feed in the developing countries has been growing at a rate of 4.6% annually, and this trend is likely to continue, and even accelerate, in the future.

During the enclosure of the English commons the dispossessed commoners used the slogan 'Sheep eat men'. A similar complaint could be made today, as cattle, pigs, poultry and even fish eat an increasing fraction of our staple foods. But the great majority of the world's under-nourished poor live in countries where food of animal origin constitutes a very small proportion of the average diet, where food production is already not keeping up with population growth, and where the ability to import and distribute food to the undernourished is very limited. Feeding grain to livestock in other countries should not be blamed for the scale of malnutrition in the world, a problem requiring other solutions. The greatest problem posed by the rapid increase in grain feeding of livestock is that grain production must not only match the increase in population but substantially exceed it to provide the requisite animal feed as well as food, thereby putting additional pressure on the resources and sustainability of agriculture.

## 11.6 Food trade, aid and stocks

With food grains, especially cereals, as with many other aspects of life, 'there is a tide in the affairs of men', as Shakespeare put it. In times past the tide of grain flowed from the colonies to the capitals of empire, from Egypt and North Africa to Rome, from Canada and Australia to London. The flow continued in that direction until World War II, but since then it has reversed direction, flowing with increasing momentum from the more to the less developed countries, as illustrated in Figure 30. Even today, however, the proportion of world food entering international trade is small. Most is for local consumption.

Although the developing countries comprise 77% of the world population, they produce only 62% of its supply of cereals. At present the net export of grain from the more to the less developed countries represents 15% of the former's production of cereals and 9% of the latter's consumption, and is primarily used for animal feed. This tide of grain is projected to flow even more strongly by 2020 AD, with one sixth

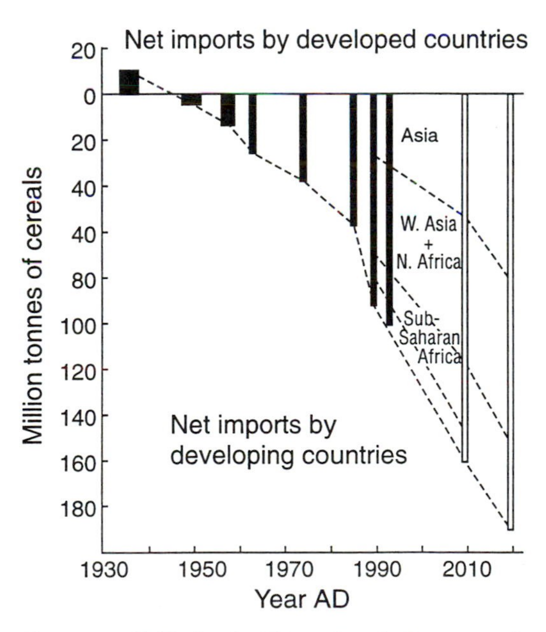


**Figure 30** Shifts in the flow of grain between developing and developed countries, together with projections by Rosegrant *et al.*[175] Data from FAO Trade Yearbooks.

of the production in developed countries furnishing one eighth of the consumption in developing ones[175].

However, the net flow of cereals to the developing countries represents less than half of the total world trade in grains between countries, which is about 12% of total grain production. The extent to which they are traded internationally varies greatly between crops. Wheat is by far the dominant foodstuff traded, with more than one fifth of its total production in recent years moving internationally, mostly to Asia and Africa. By contrast, only 2–3% of world rice production is traded internationally, mostly within Asia and to Africa and Europe. Local self-sufficiency is therefore extremely important in rice production. About 13% of the production of the coarse grain cereals, maize and barley, is traded internationally, mostly for animal feed and mostly to Asia. International trade involves about 11% of pulse grain production, 2.5% of potatoes and 27% of sugar. For the various meats, the proportion traded internationally is 4–9%. The important conclusion is that, despite the increasing dependence of developing countries on imports of wheat and feed grains, the adequacy of food supplies remains largely dependent on national and local production, and on their ability to keep up with population growth.

When famines strike, food aid can be of crucial significance, provided it actually reaches the starving people. But even in 1992/3, when shipments of food aid in cereals reached their peak to date of 17 M tonnes, this represented only 2.1% of production by the developed countries, and only 1.4% of that by developing ones, but 15% of the net flow between them. Sub-Saharan Africa is still the major recipient region for food aid, while Egypt and Bangladesh are the main recipient countries. Besides its role in emergencies, food aid has been used in attempts to enhance development, so far without much success.

Changes in the end of year world stocks of grain appear in the news whenever they get lower than usual. Prophets of doom use them as if they are a barometer of world food-producing capacity. In *State of the World 1995*, for example, Lester Brown warns that 'with carryover stocks at such a low level [then equivalent to 59 days of consumption], the world is now only one poor harvest away from chaos in world grain markets'. Grain stocks subsequently fell to only 47 days (13% of annual consumption), but the world has not had to tighten its belt. Certainly, poor harvests in some regions may have an impact on food stocks, but lack of economic incentive to produce more has a far greater effect.

The important concern when grain stocks reach low levels is not that the world is running out of grain but that the rise in prices needed to provide the incentive for farmers to restore stocks to a higher level will put food out of reach of many of the poor in developing countries. As the governments in developed countries, which hold the greater part of grain stocks, reduce their farm price support programmes with the implementation of the Uruguay round agreement, their grain reserves may be reduced. This may cause stocks and prices to become more volatile, offering yet more scope to doomsters and hurting the already malnourished poor.

## 11.7 Some projections into the future

Many projections of future world food supplies tell us more about the innate optimism or pessimism of the projectors, as expressed in their critical assumptions, than about what will actually happen. Moreover, the uncertainty principle may operate in that what eventually does happen may be influenced by policies adopted on the basis of economic projections.

The demand projections are usually driven by such factors as population growth (the most predictable element), changes in income and the inequalities of its distribution, urbanization (which can have a pronounced effect on diet), and market development. Although these, especially changes in income, may be difficult enough to predict, the supply-side factors are even more so through their short-term dependence on weather and prices and their longer-term dependence on the supply of fertilizers and other inputs and on investments in irrigation, agricultural research and infrastructure.

Who in 1960 would have foreseen that the combination of dwarf cereals, cheap nitrogenous fertilizers, new herbicides and investment in irrigation would have such prolonged and synergistic effects on food production? And who, now that the impact of that combination is dwindling, can know what rate of improvement in crop yields to assign to various regions over the next 20–50 years? Yet small differences in these assumptions can lead to projections with very different consequences over such time intervals.

The world population will reach seven billion close to the year 2010, for which there are several independent projections of global and regional cereal demand and supply prospects[1,2,138]. Comparison of these scenarios is instructive. On a world scale their projections of demand in AD 2010 differ by only 4%, with an overall growth rate in production of 1.5–1.7% per year, less than the 2.2% for the 1970–1990 period. Substantial differences between the projections begin to appear in the estimates of production by the developed and the developing country segments, and therefore of the extent of net trade between them[103]. The FAO study[2] is based on a comprehensive evaluation of land and water resources in the developing countries. It envisages the fastest increase in their cereal production among the three projections, 2.1% per year (cf. 3.1% and 2.7% in the two preceding decades), plus net imports by the developing countries of 162 M tonnes per year. Agcaoili and Rosegrant of IFPRI[1] project a similar level of imports despite their estimate of 16% more production by the developed countries and 7% less production by the developing ones than in the FAO projection. Mitchell and Ingco of the World Bank[138] project lower production by the developed countries, as in the FAO study, but also a slower increase in production by the developing ones, requiring a massive rise in their annual cereal imports to 210 M tonnes.

Although cereal production in the Middle East/North Africa is projected to grow almost as fast as population, that region will remain by far the greatest importer of grain in all projections, accounting for 34–44% of the net import of cereals by developing countries, thanks to oil export revenues, while they last! Latin America in recent years has accounted for about 12% of the cereal trade, which will increase somewhat because increases in production are projected not to keep up with the rise in population and standard of living. South Asia sees the three models in reasonable agreement that production will increase by 1.7–1.9% per year, substantially less than the 2.7–2.9% in the preceding decades, and less than the projected rise in consumption of 2.0–2.2% per year. Nevertheless, although imports increase, they remain only 3–10% of consumption. For China, the IFPRI projection assumes a substantially lower increase in production (1.3% per year cf. 1.9% in the other projections and 3.0 and 4.0% in the preceding decades), but continuing reliance on local production. Net imports would account for only 3–5% of consumption by China, compared with 11–19% in projections for the rest of east Asia and the Pacific.

The three projections diverge most sharply on what will happen in sub-Saharan Africa. All agree that production will not keep up with consumption and that net imports will increase 2–6-fold. The FAO study projects a high rate of increase in production, 3.5% per year, which is rather higher than the rate of increase in population and a triumph of hope over experience. In the other two projections, however, production does not keep pace with population. In the IFPRI model, the four-fold increase in cereal imports allows consumption to increase, leaving open the question of how they will be financed, but in the World Bank study this does not happen and the rate of increase in consumption (2.0%) is far below that in population. The chilling implications of that projection for the region of the world which already has substantially the lowest calorie and protein intake per day are a challenge to our humanity. Such projections are not predictions, however, and the differences between the three projections highlight the problems of looking ahead even to a population of seven billion, let alone to eight or ten billion.

The world's population will reach eight billion close to AD 2020, for which the IFPRI group has published projections[175], as has Dyson[56]. By then the population of China is expected to be 1.5 billion, of India 1.3 billion, and sub-Saharan Africa 1.1 billion. Of the world population then, 82% will live in developing countries. Besides their standard model, the IFPRI group also made projections for alternatives with slower population growth, trade liberalization, and with more or less investment in agricultural research, which has a striking impact on the projections through its effects on crop yields. But the main concern of the authors is that the tide in the flow of grain will still not have turned, and that the contrast between the two worlds, the haves and the have-nots, will remain almost as strong as it is now, with little improvement in the food security of many developing countries.

CHAPTER 12

# Feeding the ten billion

## 12.1 Introduction: routes to greater food production

The medium projection of world population by the United Nations in 1997 forecasts that it will reach 9.4 billion by AD 2050 whereas the high variant reaches 10 billion by about 2040. By 2050 the total population of the more developed countries will be falling. Among the developing countries, those whose populations will have increased the most between 1995 and 2050 will be India (by 575 M), China (by 301 M), Nigeria (by 208 M), Pakistan (by 205 M) and Ethiopia (by 142 M)[63]. By 2050, on the medium projection, there will be only 1.2 billion people in the more developed regions compared with 8.2 billion in the currently less developed, 5.4 billion of them in Asia and 2 billion in Africa. This population explosion in the developing countries is historically unique, and matching the expansion of the food supply to that of humanity will be a singular challenge.

The world must develop the capacity to feed the ten billion within the next 40–50 years, predominantly within Asia and Africa. In this context it is important to distinguish two quite separate but often conflated problems. The first is that of developing the global capacity to produce enough food for ten billion people, i.e. for 67% more than at present, the main focus of this book. The second is that of eliminating the chronic undernutrition which still afflicts so many of us in a world which produces enough food for all. Agricultural research is the key to the first problem but cannot be expected to solve the second, more complex, task of eliminating poverty and providing the work, health and education which should allow the poor to obtain food.

The six main components of increased global food supply by crops are:

(1) increase in the area of land under cultivation;
(2) increase in yield per hectare per crop;
(3) increase in the number of crops per hectare per year;
(4) displacement of lower yielding crops by higher yielding ones;
(5) reduction of post-harvest losses;
(6) reduced use as feed for animals.

Several of these are considered in more detail later. Increase in the arable area was the dominant component of increases in food supplies until the world population reached three billion in 1960. Although new land is still being brought into cultivation, particularly in South America and Africa, these gains have been balanced over the last 30 years by losses to the spread of cities and infrastructure and to erosion and degradation. The rural population of the world is expected to remain about 3 billion[56], so nearly all the population growth will be accommodated in cities, which are often located on the best agricultural land. Although there is still considerable scope for conversion to arable land, especially in South America and Africa, pressures for the conservation of nature and of biodiversity will limit this option.

Since 1960, increase in yield per crop has been the dominant component of greater food production (Figure 17, p. 91), to such an extent that the relation between average cereal yield and world population has been very close since 1960. This has already saved much land from the plough, and will continue to do so for some time, but high input agriculture inevitably generates new problems and concerns.

The other form of intensification, particularly in the tropics, is increase in the number of crops per year from a given piece of arable land. Mostly this requires access to irrigation as well as high input use and the breeding of short season varieties. The additional requirements for labour with this form of intensification, and greater pest and disease pressures, can be problems.

The displacement of less productive by more productive crops has been going on since the beginning of domestication. The advent of cheap nitrogenous fertilizers disadvantaged the legumes *vis-à-vis* the cereals, which also benefitted from the introduction of herbicides. However, the present dominance of the cereals may, once the requisite food supply is assured, be reduced, particularly by the desire for more varied diets.

The scope for reducing post-harvest losses remains unclear. Pimentel[156] estimated such losses to be about 20% on a world scale, ranging from 9% in the USA to 40–50% in some developing countries. However, losses during storage in several traditional systems may be very low, as little as 2%. Greeley[75] believes the losses in developing countries have been exaggerated, creating what he calls 'the myth of the soft third option' for solving the food supply problems of developing countries simply by reducing post-harvest losses.

By contrast, a radical change in the consumption of animal products in our diets could have a huge impact on the amount of grain available *globally* for direct human consumption. One estimate for 1980 suggested that 44% of the world cereal production was being used for animal feed, another for 1982–89 gave a comparable figure. Although that proportion may have fallen slightly in recent years, it seems more likely to rise in the longer term as animal products become more prominent in the diets of developing countries.

Thus, further increases in yield per crop seem likely to provide the main route to feeding the ten billion, although enhancing yields at the lower latitudes in the context of likely changes in climate will be a major challenge. Raising yields fast enough to keep pace with the rising demands for food and feed for the world as a whole will be sufficiently difficult that, except perhaps to a limited extent in some developed countries, there will be little scope for crops to provide sources of energy other than those for human and animal metabolism. The following sections consider some of the factors bearing on these routes to greater food production.

## 12.2 Will there be enough arable land?

Until the world population reached three billion, extension of the area under cultivation was the main source of greater global food production. Then, quite suddenly, that expansion slowed and the present area of arable land (Figure 31), 1.34 billion hectares, about 11% of the land surface not covered by ice, is only slightly greater than it was in 1960. However, the basis of the FAO statistics has changed somewhat over the years. Until the 1970s the forest fallow areas associated with shifting cultivation were classified as arable and when this convention was discontinued the arable areas of Africa and South America appeared to

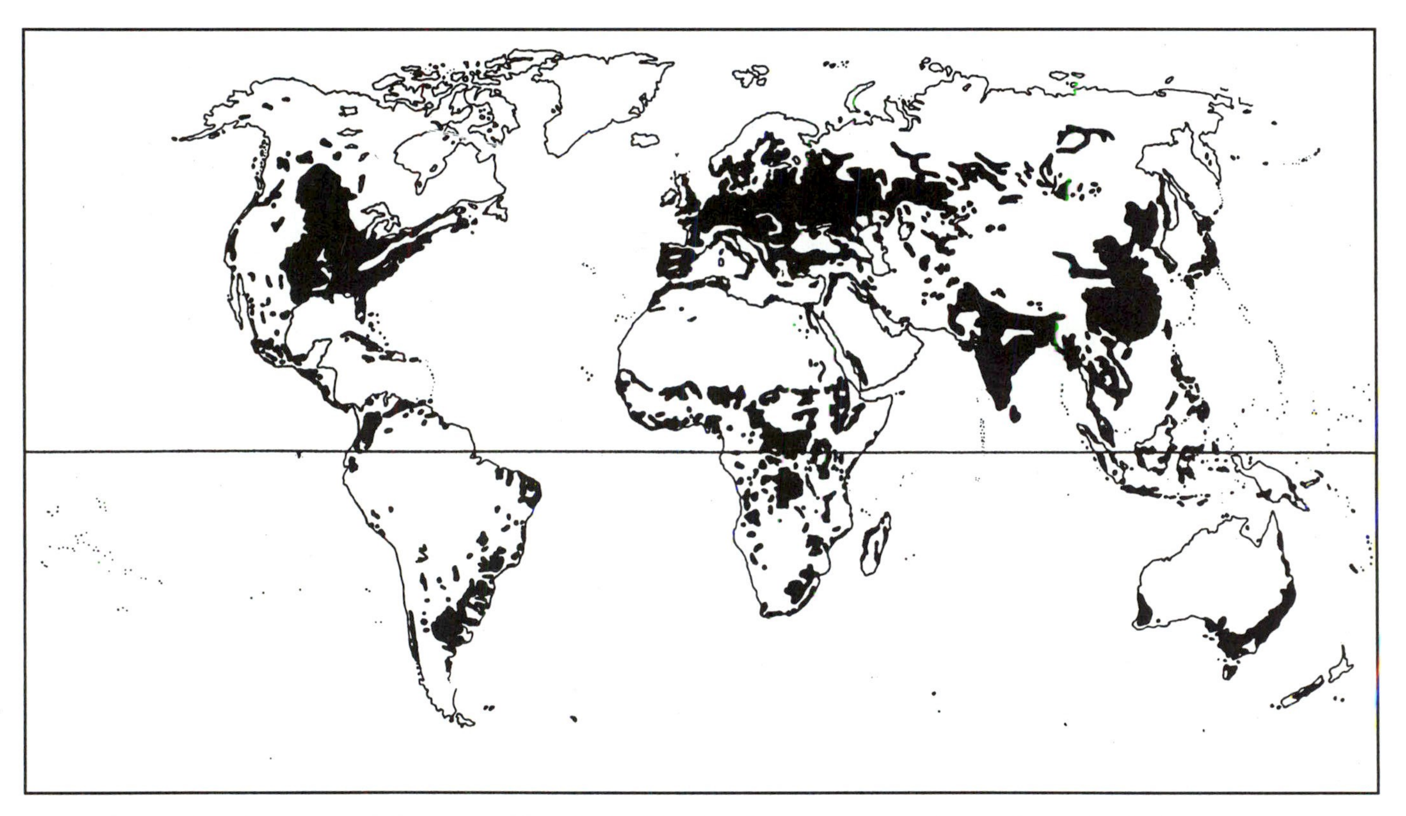

**Figure 31** The distribution of permanently cultivated land. (Adapted from the Times Atlas of the World.)

decline by about 9% in both cases. With double cropping in Asia the arable area used to be counted twice, again resulting in an apparent decrease in arable area when that convention was changed.

The recent stasis in arable area is not because the world has run out of potentially arable land. Far from it. Several independent estimates indicate that between 3.0 and 3.4 billion hectares (ha) of land are potentially cultivable[25,164]. Much of the presently uncultivated area is already used for grazing livestock or is of poorer quality, too remote or subdivided to be economic, vulnerable to erosion or cherished in its present state. Fischer and Heilig[63] estimate that there are 2.5 billion hectares suitable for crop production in developing countries, of which only about a third is currently in cultivation. However, the balance available for rainfed cropping shrinks to only 550 M ha when all forest and wetland areas are excluded from agricultural use.

In the developed countries, which have almost half (48%) of the world's arable land but less than a quarter of the population, 77% of the potentially arable land is already cultivated. In the developing countries this proportion is only 36% but ranges from 15% in South America and 21% in Africa to over 90% in Asia[25]. However, more land may be cultivated in some developing countries than their FAO statistics indicate and there may be less potentially cultivable land than the figures above suggest.

Why did expansion of the arable area stop in the 1960s? One apparent answer at the global level, although not true of Africa, can be seen in Figure 17 (p. 91). Once the rise in the average world yields of the staple cereals began to keep pace with population there was less pressure for further extension of the arable area.

The crucial role of higher input agriculture in making possible the conservation of otherwise threatened natural communities is not always appreciated. In 1977 Buringh and van Heemst[26] estimated the human carrying capacity of the world if only traditional subsistence, labour-oriented agriculture was used on *all* the land suitable for such farming. They found it would be impossible to feed even the four billion people then alive, in spite of the huge increase in land to be cleared for agriculture. So the intensification of agriculture and the raising of yields was essential, as well as timely, besides saving about 1.8 billion ha from the plough. Much of the continuing deforestation is driven by the demand for timber rather than for food. Although estimates of the extent of recent tropical deforestation vary substantially[134], even the highest is equivalent to less than 1% increase in the arable area each year if all the cleared land were converted to agriculture. This is certainly not the case, although much of it may end up slashed and burnt for a time.

As crop yields are raised by the intensification of agriculture, the amount of arable land needed to supply each person falls. In 1960 there was 0.47 ha of arable per head for the world as a whole. There is now only half as much and the world is, on average, better fed. In the developing countries the current average is 0.16 ha per head, and only 0.08 ha in Bangladesh and China, yet the latter is largely self-sufficient for its food. Egypt, with its heavy reliance on irrigation by the Nile, as well as on food aid, currently has only 0.04 ha of arable land per head.

Illustrations showing the fall in arable land per head are sometimes used to bolster the claim that the world is running out of land for food production, whereas what they really show is how much less arable land per head is needed as the intensification of agriculture proceeds. Nevertheless, the apparent stability of the area under cultivation on a world scale conceals some disturbing trends in both the losses and the compensating gains.

The most serious loss is that to urbanization (i.e. village expansion, urban sprawl, roads, airports, industry, etc.) which now exceeds, and occupies much of, what was the arable area of the world in AD 1700. Buringh and Dudal[25] estimate that by AD 2000 about 400 M ha of mainly prime land will be used for non-agricultural purposes, which will be equivalent to 0.13 ha per urbanite, i.e. four fifths as much as the average amount of arable land per head used for food production in developing countries. In them, however, the average area for habitation and infrastructure may be closer to 0.05 ha per person currently[215]. The proportion of urban dwellers in developing countries by the end of the millennium will be 44%, twice as great as it was in 1960, and will reach 57% by 2025. Moreover, cities tend to grow on the best agricultural land. One estimate indicates that although only 4% of all potentially agricultural land will be lost to urbanization between 1975 and 2000, that will include a quarter of the most highly productive land[25]. 'Urban agriculture', despite the apparent oxymoron, merits more attention and quantification.

When we turn to losses of arable land by erosion, desertification, salination, toxification and other forms of mismanagement, the many estimates vary wildly. Much depends on quantifying the rate as well as the extent of losses. Most arable lands, apart from rice paddies, lose some soil each year, and many irrigated soils show some salting, without loss of yield. More severe degradation may cause loss of yield but not the irreversible loss of arable land. Extreme gully erosion, salination, toxification or the desertification of arid land to the stage of irreversibility occurs on a far smaller scale. Buringh and Dudal estimate that each year since 1975 about 2 M ha of arable land have been seriously toxified and another 2 M ha desertified compared with about 8 M hectares converted to non-agricultural (mainly urban) use. The total of 12 M hectares per year represents a loss of almost 1% of arable land each year.

Much of the arable land lost to cultivation because of salination would not be usable without irrigation anyway, and much of that lost to erosion is in drier areas which are marginal for cropping, so the losses to urbanization, being mostly on prime agricultural land, are by far the most serious. Although this is being replaced by newly cultivated land, the reserve of potentially productive agricultural land from grazed grasslands and forests is dwindling.

On top of these losses of land from cultivation there are also the losses in the productive capacity of the remaining arable land. One recent survey suggests that 38% of our cropland has been degraded to some extent over the last 50 years[185], another that 47% of the rainfed drylands and 30% of irrigated drylands have been degraded. Degradation is most extensive in Africa (65%) and Latin America (51%), the two regions where, fortunately, the proportion of potentially arable land which is currently cultivated is lowest (only 21 and 15% respectively). Indeed, these are the regions where the area of arable land is still increasing, offsetting the small declines in Europe and North America.

In projections forward to AD 2010, further increases of about 30% in the arable area of sub-Saharan Africa and Latin America are anticipated by FAO, compared with only 4% in south Asia and 9% in the Near East/North Africa region[2]. Thus increase in the arable area is expected to play a relatively minor role in the further expansion of world food production, but significant extension of the arable to less favoured soils and environments will be needed to replace urbanized or degraded areas. The conservation and improvement of already cultivated soils should rank high in our priorities, but is not always feasible or economic when farms are small or grain prices low.

## 12.3 Intensification

Ester Boserup's thesis that population pressure drives the intensification of agriculture grew out of her observation of the progressive shortening of the fallow period in shifting cultivation from forest (20–25 years) to bush (6–10 years) to grass (1–2 years) to annual and then multiple cropping. Such progressive intensification in response to rising population pressure occurred in Europe and China in earlier times and is still to be seen in Africa, South America and Asia.

Where soil fertility cannot be maintained by means other than the fallow, yields and the returns on labour will decline with intensification. As this proceeds, greater nutrient input is required and, once the stage of grazing animals on pasture leys as a source of farmyard manure is passed, this must come from legume crops in a rotation or from fertilizers. Still further intensification requires still heavier applications of progressively more nutrients to replace those removed by the crops.

More crops also require more water, so supplementary and eventually full irrigation may be needed. As cropping becomes more intensive and the diversity of crops is reduced, weed, pest and disease problems become more acute, requiring further inputs. More labour is also needed to tend the crops, which can provide more opportunities for the poor and landless to earn income, and raise real wages even in densely populated, labour-surplus countries such as India. However, the incentives to still further intensification may be lessened for the farmers.

Of the currently arable land in developing countries, only in Asia, both south and east, does the average intensity of cropping (i.e. area in crop/arable area) currently exceed more than one crop per year. It ranges from 55% in sub-Saharan Africa and 61% in Latin America to 83% in the Near East/North African region. Much of the uncropped arable land is in the drier areas where fallowing is needed to gain sufficient moisture for another crop, and cropping intensity in these areas exceeds 100% only when irrigation is available. Thus, there may be little scope for fuller use of the already arable land in developing countries except where irrigation is possible.

Double cropping of rice has been known in China since Sung times (12th century AD) and has spread progressively northwards. The key to it was the introduction of the early flowering Champa rices, together with irrigation and transplanting. In much of China and other irrigated areas such as the Punjab, the cropping intensity now approaches an average of 2, and could go still higher as even earlier maturing varieties with high yield potential are bred. Rice is particularly well suited to it and one sixth of the global area under rice is double cropped[76]. More diverse systems, such as wheat–rice rotations, have been developed, and further improvements in tillage, fertilizer use and varietal characteristics – especially early maturity – could raise the cropping intensity still further. Alexandratos[2] estimates that of the increase in crop production by developing countries between 1988–90 and 2010, two thirds will come from rising yields per crop, 20% from extension of the arable and only 13% from greater cropping intensity. The availability of irrigation water may prove to be the limiting factor to intensification.

At the International Rice Research Institute in the Philippines three crops of rice per year have been grown on one field without interruption for 34 years. The hundredth crop was planted in 1997 immediately after its high-yielding predecessor was harvested, indicating that intensification need not compromise sustainability. A comparison of the cumulative yield of rice from the *unfertilized* plots at IRRI, 288 tonnes $ha^{-1}$ in 30 years, with those of wheat from unfertilized plots at Rothamsted, 210 tonnes $ha^{-1}$ in 140 years, highlights the potential of multiple cropping in favourable areas of the tropics.

## 12.4 The imperative of further increase in yield

Increase in the average yields of the cereal staples has largely kept pace with the increase in world population since 1960 (Figure 32) and it is widely assumed that it will continue to do so until the population begins to stabilize in the 21st century. Between 1966 and 1990 the rate of increase in grain *production* was 1.87% per year and increase in yield has accounted for 92% of this since 1974[1].

Projections to AD 2010 by FAO, IFPRI and the World Bank all assume that the further increases in cereal production will come from continuing increases in yield. That seems the most likely route, given our recent history. But can a continuing and sufficient rise in yield per crop be taken for granted as so many economists and demographers seem to assume? A glance at Figure 32 will show that the average world cereal yield would have to reach at least 4.0 tonnes $ha^{-1}$ for a population of eight billion, and 5.0 tonnes $ha^{-1}$ for a population of ten billion, from its

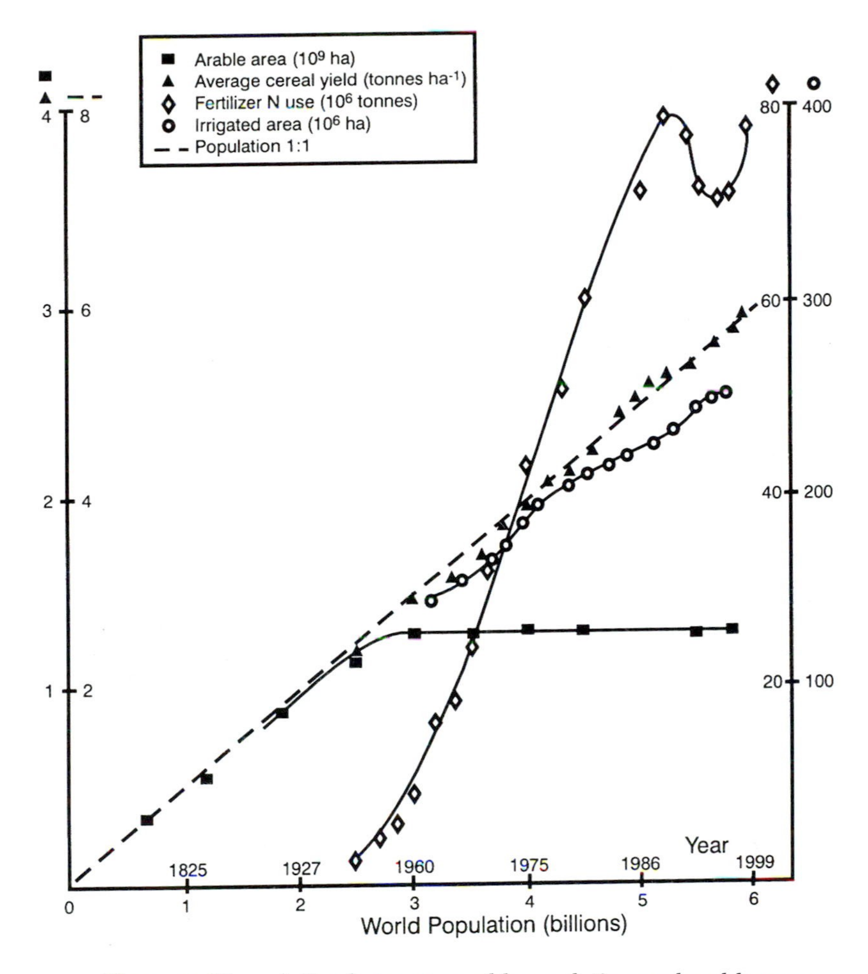


**Figure 32** The relation between world population and arable area, average cereal yield, N fertilizer use and irrigated area. (Data from FAO Production Yearbooks.)

present level of 3 tonnes $ha^{-1}$. Even for eight billion people, the *average* world yield would have to equal those of Europe and North America today, and exceed them by more than 25% to sustain ten billion.

Undoubtedly there is still some momentum, at least in many developing countries, for further increases in yield from the innovations of the Green Revolution. Although more than 200 kg of N fertilizer per hectare of arable land are already being used in Egypt, China, the

Indian Punjab and the Yaqui Valley of Mexico, the average use is much lower and still increasing, except in Europe.

The earlier chapters of this book have shown us over and over again that there are no grounds for assuming there will not be further advances in yield just because we cannot foresee a route to them. Nevertheless, the conjunction of three innovations (cheap nitrogenous fertilizers, dwarf cereals and improved weed control) which gave the Green Revolution such a timely and potent impact on world food production, may prove to be unique. The dwarfing, made necessary by cheap nitrogenous fertilizers and made possible by herbicides, had a double impact on yield. By reducing the lodging of cereals it made heavier dressings of fertilizer worthwhile. And by reducing the requirements for stem growth, it allowed more investment in the development of the seed head and grain growth, thereby raising the harvest index and the yield potential.

Further increases in the harvest index will be limited and other sources of greater yield potential, such as faster photosynthesis or growth, do not seem imminent. However, crop yields could go on increasing to a significant extent even in the absence of any further rises in yield potential, for several reasons. The yield potential is the yield of a cultivar when it is grown in environments to which it is adapted, with nutrients and water non-limiting, and with pests, diseases, weeds, lodging and other stresses effectively controlled. Actual yields are mostly far below that potential because of imperfect adaptation to local environments, insufficient provision of nutrients and water, and incomplete control of pests, diseases and weeds. There is also considerable scope for further investment in land improvement through drainage, terracing, control of acidification, etc. where these have not already been introduced. Such investments have often been subsidized by governments in developed countries, and may require targeted aid to developing countries.

Thus, yields may well be raised even if yield potential is not. But whether they can be doubled, on average, over the next 50 years or so without further raising the yield potential is open to question. With wheat in the United Kingdom, dwarfing has been associated with a 60% rise in yield potential, which has contributed about half of the increase in average yield. With maize in the USA, on the other hand, although plant breeding has contributed over half of the rise in yield, its contribution has been mainly to greater tolerance of closer planting, environmental stresses, pests and diseases rather than to yield potential. Height has not been reduced and the harvest index has not been raised markedly.

Clearly there are several routes to higher yields, even among the cereals, and they may be taken indirectly and unconsciously in the process of empirical selection for higher yield. For example, maize varieties selected for yield at high latitudes may display greater tolerance of cool nights in their photosynthesis. Similarly, both cotton and wheat varieties selected for high yield in irrigated hot, dry environments have been found to have more open stomata and cooler canopies in the afternoon.

It is the power of empirical selection for yield to take advantage of subtle environmental adaptations such as these, and to integrate them into higher yielding genomes, that offers hope that the yield of our staple crops will continue to rise in step with world population to feed the ten billion. Although both agricultural scientists and farmers find it difficult to believe that average crop yields at some time in the future will equal current record yields, that is what has happened in the past, e.g. for the maize and soybean growers of Iowa after a lag of about 50 years[58].

## 12.5 The resources for future food production

René Dubos once said, provocatively: 'There are no resources, only human resourcefulness.' The crucial role of further rises in crop yields in the absence of further expansion of the arable has just been highlighted. Such rises will come partly from plant breeding and partly from agronomic improvements which will depend, in turn, on the greater use of other resources as substitutes for land. The most important of these are irrigation water, fertilizer elements and sources of energy, and it is these that may eventually limit the world's capacity for food production.

Although less than one fifth of the world's arable land is irrigated, that small fraction has contributed most of the increase in wheat and rice production by developing countries in recent years. As the key to higher yields and greater cropping intensity in developing countries, the further extension of irrigation is essential, but likely to be limited, for several reasons.

Firstly, the demands of industry and urban areas for water are growing rapidly, and they are better able to pay for it. By AD 2000 industry is projected to account for 41%, and irrigation for only 51%, of world water use. Secondly, much of the most readily available water and land suitable for irrigation has already been tapped and the cost of new irrigation developments is rising. Thirdly, irrigation has been given a poor image in recent years, with too much emphasis on its environmental problems, from salination to schistosomiasis, and too little on its crucial role in feeding the world. It has also been argued that we humans are already appropriating too large a share of the world's annual supply of fresh water[161]. The control of water resources may well prove to be a defining issue for the 21st century.

There is substantial under-investment in new irrigation schemes and in the rehabilitation of existing ones, and too little recognition of their secondary benefits[174]. One estimate of potentially irrigable land[27] suggests that there are 470 M hectares, i.e. about twice the present irrigated area, whereas another suggests that the available water would limit the increase to only about one third[46].

The supply of fresh water for irrigation could be augmented if a source of cheap energy for desalination of seawater and pumping and distribution inland is developed. Cheap energy supplies could also make it feasible to extract the requisite supplies of mineral nutrients from lower grade ores than it is currently economic to work. Nitrogenous fertilizers could be produced in non-limiting amounts, as could the requisite pesticides and other agrichemicals. The scale of industrial nitrogen fixation is already comparable with that by the earth's terrestrial biosphere and will exceed it before there are ten billion of us. Supplies of phosphorus and potassium should also be ample.

We have already seen how human labour and animal power have been replaced and augmented in modern agronomy by mechanization, electrification and the industrial production of an increasing variety of inputs. Energy supply will continue to be a major component of future agricultural development and may become the primary limitation to it insofar as it is the key to enhancing resources of water, arable land, mineral nutrients and other inputs. As Vasey[212] puts it: 'Food futures hinge on population and energy futures'. Agriculture currently consumes only about 5% of the world's energy use, nearly all of it based on non-renewable fossil fuel. In that respect modern agriculture is not as sustainable as it was until there were only two billion of us.

Repetto[167] argues that agriculture will be truly sustainable only when it relies on Nature's income, not on its capital. While there may be limited scope for liquid fuel crops, agriculture could not possibly produce both the input energy it needs and the food needed by the ten billion. Although agriculture cannot achieve Repetto's goal, however, more than 99% of the total energy intercepted by or applied to high yielding maize crops still comes directly from the sun, not from fossil fuels. What the other inputs achieve is a greater efficiency in converting that intercepted solar energy into food.

## 12.5 Old and new crops

We are an exploratory, inventive species, fascinated by the new, dispensing with the old. So 'new' crops attract optimistic headlines while old ones are often blamed for the problems of agriculture. Yet the world still relies on three of the oldest domesticates – one from the Near East, one from Asia and one from the Americas – for the bulk of its food and feed supply.

More than two thousand plant species have been domesticated to some extent over the last ten thousand years. Many of these are still cultivated on a small scale, often quite locally, like teff in Ethiopia. Others have been either replaced or displaced to poorer conditions by crops which are more productive, reliable, adaptable or nutritious, and the most important of these have become the staple crops which provide most of the world's food.

Since the introduction of cheaper nitrogenous fertilizers the cereals have extended their dominance of our food supply, at the expense of legumes and root crops. Likewise, wheat, rice and maize have become progressively more predominant among the cereals, their share of world cereal production having increased from 69% in 1950 to 84% today. They dominate the diets of many developing countries and the food and feed supplies of developed ones. But dependence and blame are frequent fellow travellers, and intensive cereal cropping is often faulted for being less sustainable and more environmentally harmful than, for example, legume cropping.

New crops have been promoted on the grounds that there is insufficient genetic diversity in the old ones, despite the comprehensive gene bank collections of wild relatives, land races and varieties of the major crops, e.g. of more than 80,000 accessions for rice. Modern varieties of the staples are also claimed to have narrow genetic backgrounds and therefore to be vulnerable to pests and diseases. But with almost 70 land races, from many regions, in the ancestry of recent CIMMYT wheat varieties and a comparable diversity in dwarf rice and hybrid maize ancestries, their genetic background has never been so wide[59]. Any variety outstanding enough to be grown on 10 M ha must have a comprehensive set of genetic resistances and will have those exposed to extensive challenge by pests and diseases. Narrow ecological adaptation is another brickbat thrown at the staple crops, yet one of their outstanding features and a key to their success has been their adaptability. Nor do they 'need' high inputs, as is often claimed. Rather, they respond well to them.

So-called new crops fall into several categories. Most are old crops in the sense of having been domesticated long ago, often cherished for the variety they added to diets and farming systems but gradually displaced by more reliable crops (e.g. Job's tears), by religious decree (*Amaranthus* spp.), by mechanization (e.g. buckwheat, the preferred porridge of Anton Chekhov), etc. There are many books on these, such as *The Lost Crops of the Incas* (1989).

A second category of 'new' crops consists of old ones put to new uses, as when fodder beet (chard) was selected as a source of sugar in Napoleonic times. Since then many other crops have been selected to enhance or reduce particular constituents for new purposes. Canola and linola are recent examples. Canola was developed from oilseed rape, firstly by virtually eliminating the undesirable erucic acid and largely replacing it with oleic acid. It now ranks third behind soybean and palm as a source of vegetable oil, and the only one adapted to cool temperatures. Secondly, by greatly reducing the undesirable glucosinolates in the residual meal after oil extraction, canola also became an acceptable feed grain.

Other examples could be given and many more are in the pipeline for the future, based on genetic engineering to enhance or inhibit specific constituents for nutritional or industrial purposes. Prime targets for such biotech breeding are the structure and composition of stored starches and lipids in readily transformable crop plants. Besides expanding the diversity of potentially profitable agricultural crops, such biotech crops could provide renewable sources of feedstocks and reduce toxic waste problems. Intensely sweet proteins like monellin and thaumatin (a hundred thousand times sweeter than sucrose), antibodies and other pharmaceutical compounds are also likely to be produced in the future by old crops transformed for new uses.

In the truly new crop category the species range through oil crops such as jojoba and *Melaleuca*, still not domesticated; American wild rice (*Zizania*), still in the throes of being domesticated; several recently domesticated lupin species now used as feed grains; and triticale, on the way to becoming an important crop which could challenge wheat on sandy or acid soils, and in cold climates, as feed and, if its protein composition can be modified, even for bread making. There are also many suggestions for new crops based on slender ecological observations. The first entry in a book on unexploited tropical plants was a wild Australian grass called Channel millet which caught the eye because it supposedly required 'only a single watering' on its way to maturity. Of such stuff are dreams made. It happens to grow naturally on deep, fertile soils well supplied with underground water, and it turned out to have no advantages over already domesticated and closely related *Echinochloa* millets from India and Japan.

It seems unlikely that 'new' crops will reduce our dependence on rising yields of the old staple crops to match food supply with population growth. Indeed, together with further loss of much of the best arable land to urbanization, the growing of genetically engineered crops for industrial feedstocks and pharmaceuticals will enhance the need for still higher yields from the staple crops. Whether the 'new' crops are 'lost', genetically engineered or newly domesticated, they can play a valuable role in the diversification of both our diets and agriculture. Initially there may be less scope for such diversification in the developing countries until the food supply gets ahead of population growth, but in the developed countries the staple food crops may give way increasingly to those enhancing dietary variety and to 'boutique crops' such as wild rice and exotic fruits. The role of the staple cereals in feeding the world might then become less crucial.

## 12.7 Global climate change and food supply

This is not the first time that agriculture has encountered a period of climate change. Indeed, V. Gordon Childe, who fathered the concept of a Neolithic Revolution, believed it was climatic change which led mankind to the agricultural way of life. Rowan Sage[180] has suggested that it was not until the rise in atmospheric $CO_2$ level between 15 K and 12 K years ago that cropping could have been sufficiently productive to make agriculture worthwhile. Since then the decline of Sumerian irrigation (over 4 K years ago), the desertification of the Negev and the erosion of the Aegean hills over 2 K years ago, the migration of Arabs to the Near East about 700 AD, the decline of the Maya about 800 AD and 'the little Ice Age' from 1550 to 1850 have all been associated with periods of changing climate[104,178]. But with the human population currently ten times greater than it was in AD 1600, the stakes are now higher.

We know quite a lot about the growth and yield responses of crop plants to atmospheric $CO_2$ concentration and to temperature. Global weather patterns, atmospheric $CO_2$ and methane concentrations, and solar radiation and ultraviolet levels are continuously monitored. Dynamic modelling now allows national, regional and world agricultural production and even prices and international trade to be projected many years ahead, with or without various elements of climatic change.

Before considering such projections however, we should look at some of the more important elements of current change. Primary among these is the rise in the atmospheric concentration of several 'greenhouse' (i.e. heat-trapping) gases such as carbon dioxide, methane, nitrous oxide and chlorofluorocarbons as a result of industrial, agricultural and other human activities. When the heat-trapping effect of all the gases is converted to $CO_2$ equivalents, the doubling of its concentration (i.e. reaching 600 parts per million) is expected to occur by the year 2060.

As the greenhouse gases increase, so also will the average global temperature, with estimates varying between 1.5 and 4.5°C for the doubling of $CO_2$, less at low latitudes, more at high. However, the extent of this warming may be reduced by the counteracting effects of sulphate aerosols on cloud formation. Seasonal rainfall patterns are also likely to be affected, to uncertain extents. At high latitudes, in the more developed countries, warmer conditions may often be agriculturally advantageous, allowing new arable land to be brought into production and extending the season for crop growth. At low latitudes, on the other hand, crop duration and yield will be reduced but opportunities for intensification increased. With the rise in temperature there will also be a rise in sea level and some loss of arable land, e.g. in Bangladesh.

The rise in temperature will reduce the water use efficiency of crops, but that effect may be counterbalanced by the rise in $CO_2$, which tends to reduce the density and opening of the stomatal pores in leaves. However, that response could in turn raise leaf temperatures and increase heat injury. The net effect of these responses will vary from crop to crop and with the extent of cloud cover, one of the least predictable of the effects of global warming, along with changes in rainfall patterns.

On top of these counteracting effects of global warming on crop production, there are the 'direct' effects of the rise in $CO_2$ concentration[47]. For many crops, particularly under high light conditions, a rise in $CO_2$ initially results in faster photosynthesis and growth. In some plants this persists and results in increased yield. In others, reserves accumulate in the leaves, stomata close a bit, photosynthesis returns to its previous rate but transpiration is reduced and water use efficiency rises. More sugars and other assimilates may be excreted into the soil, raising its organic content and encouraging soil biota. In the longer run plants may adapt by reducing stomatal density, presumably because greater water use efficiency is often of more adaptive value than greater photosynthesis.

Clearly, the net result of rising $CO_2$ and temperature on crops is difficult to predict. Sometimes they have opposite effects, as with water use efficiency where they may cancel one another out. Sometimes they reinforce one another, e.g. in their effects on leaf temperature, since stomatal closure at high $CO_2$ may also lead to a rise in leaf temperature. Moreover their short-term effects may differ from the longer-term ones, and individual species will react differently to them. For example, plants such as maize, sorghum and sugar cane with the $C_4$ pathway of photosynthesis respond much less to a rise in $CO_2$ level than do $C_3$ plants like wheat and rice.

Only bold spirits, therefore, would try to project, on a world scale, the effects of climatic change on food production when there are ten billion of us. To provide some internal checks on their projections, Cynthia Rosenzweig, Martin Parry and their colleagues[64,176] used the climatic scenarios for a doubling of atmospheric $CO_2$ from three independent general circulation models (GCMs) at 112 sites around the world with crop growth models to represent the impact on about three quarters of world wheat, maize and soybean production and half that of rice. The impacts were assessed both with and without the *direct* effects of $CO_2$ on crop yields. As if all this was not bold enough, they went on to project international trade and price changes, as well as the effects of two arbitrary levels of agronomic adaptation.

Given the substantial differences between the results with the three GCMs – the British model being the most pessimistic – and the inadequacies of the crop models, the projections must be viewed with scepticism, but they are the best we have. Without the direct effects of $CO_2$, yields decline everywhere, but when the direct effects are included yields are projected to increase at middle and high latitudes with two of the GCMs, but to decrease substantially at low latitudes. A similar study using the same three GCMs but different crop models for rice in Asia came to similar conclusions. For what it is worth, the models suggest that agronomic adaptation could ameliorate the situation somewhat but it is clear that the adverse impact of climatic change will fall most heavily on the developing countries at low latitudes, particularly in west Asia, sub-Saharan Africa and Central America, implying a greater dependence on international trade for their staple foods.

It appears, therefore, that both plant breeders and agronomists will be stretched over the next decades not only to raise yields to keep pace with population growth and urban appropriation of the best arable land, but also to adapt the major crops and their agronomy to warmer climates, and possibly even more so to any abrupt change to the distinctly cold conditions which they could precipitate within a century, possibly accompanied by much drier tropics[24].

## 12.8 What chance a brown revolution?

The 'Green Revolution' has particularly benefitted crop production in areas with irrigation or a sufficient and reliable enough rainfall to guarantee returns on inputs such as fertilizers. The spread of modern varieties into the drier areas has been much slower and their impact on yields much less. With wheat, for example, yield gains over the traditional varieties have usually been less than 20%, often less than 10%, and none at all in harsh environments[31]. The use of fertilizer is risky in environments with low and uncertain rainfall, and access to it and other inputs is often poor. Traditional varieties may attract market price premiums of 15–20% for their grain, while their straw is valued as fodder for the livestock which are commonly an important component of the farming systems.

Can there be a 'brown revolution' within these and other constraints? Agricultural scientists have often been accused of turning their gaze to greener fields where progress is faster and more assured. Have we under-invested in research for marginal environments, as Derek Byerlee and Michael Morris[31] ask? There are several strong reasons for investing in such research, and aid donor agencies have been persistent in putting them forward:

(1) Many people already live in and depend on marginal environments. More than 700 M live in the semi-arid tropics alone, many of them in south Asia, but occupying more land in Africa. To these must be added those living in the temperate dry areas of the Near East and North Africa. The total number of people living in severely water-stressed environments, e.g. where yields are less than 40% of potential, is not easy to estimate, but Falkenmark[61] projects three billion of them by AD 2025.

(2) Many of those who live in marginal environments are among the poorest in the world.

(3) Many marginal environments are susceptible to erosion and degradation, and require sustained and skilful management if they are to retain their productivity.

While these are all persuasive arguments in favour of focusing more research on marginal environments, they are compelling only if there is evidence that real progress can be made, or deterioration halted, within the confines of existing agricultural systems, or if viable new systems can be devised. The traditional farming systems and varieties in marginal environments have been shaped by long and hard experience, and place a higher value on long-term stability than on higher yields in favourable years. Conventional agricultural research has, so far, had relatively little impact on marginal environments. Even in developed countries, the rise in yields has been slow in such environments, whether for dryland wheat in the USA and Australia or upland rice in Japan. But in developing countries the slow rise in yield has often been accompanied by rapid population growth, creating food-deficient areas, outmigration, reliance on remittances and almost inevitable soil degradation.

Robert Chambers[37] has argued that agricultural scientists need to get away from 'whatever is capital-intensive, mechanical, chemical, and quantifiable' to 'whatever is labour-intensive, powered by animals or people, organic and difficult to quantify', i.e. the resources of the rural poor, and to do this in consultation with 'those who are most powerless, most scattered, most unable to articulate their needs, most unable to make demands on the system'. However, given the long evolution of traditional farming systems in such environments, there may be little scope for improvement without the introduction of new elements. Seedbed preparation and the control of soil erosion and weeds might be improved in the light of modern knowledge, and water conservation enhanced. The range of crops might be expanded and farming systems modified, particularly to make more effective use of N-fixation by legumes and other methods of maintaining or increasing the amount of organic matter in the soil, including agroforestry systems. New varieties with a more comprehensive and contemporary range of pest and disease resistances might also be introduced without fundamental change to existing farming systems.

Salvatore Ceccarelli[36], who breeds barley for temperate dry areas, believes the main reason for lack of breeding progress in marginal environments is the use of principles and practices focused on favourable environments. These offer faster progress, synergistic interactions with agronomic inputs, and more widespread use of the new varieties. By contrast, selection in marginal environments offers miniscule differentials, minimum interactions with inputs and the likelihood of slightly higher yields only in some environments, which may, nevertheless, be valuable[191]. In the less limiting environments, some fertilizer use may be worthwhile, but farmers are reluctant to use it in most marginal environments because in dry years they will lose their investment, can not afford it anyway and often as not can not get it.

Nevertheless, improved varieties are the most likely Trojan horse for improved agronomy and yield in such environments. In their book *New Seeds for Poor People*, Lipton and Longhurst[119] have tried to identify useful targets for plant breeding programmes. Others have added features, such as increased mineral and vitamin contents to raise nutritional value, and apomictic (i.e. non-sexual) reproduction to fix all these together, but selection for these would only compromise the already difficult-enough task of selection for higher yield in adverse environments.

'Research is the art of the soluble', as Peter Medawar emphasized, going on to say that the spectacle of a scientist locked in combat with a problem is not an inspiring one if, in the outcome, the scientist is routed. The arguments in favour of research towards a brown revolution are strong, but until we can see a way ahead, the rapidly growing population of marginal environments will have to depend increasingly on food from elsewhere or on migration elsewhere. Byerlee and Morris[31] conclude that in the case of wheat the marginal environments are receiving, at least, their proper share of research attention. Nevertheless, there remains a real need for the development of more sustainable farming systems for the less-favoured areas.

## 12.9 Alternatives in agriculture

The high input agriculture on which the world now depends for its food supply has many critics, across a wide spectrum of dissatisfaction. Some hanker for the old self-reliant family farms. Some are apprehensive about the long-term effects of pesticides, herbicides, antibiotics and fertilizers on human health; others about the effects of erosion, both of soil and of genes, on environmental health and biodiversity. Yet others seek a sustainable or even a self-sustaining agriculture with a long-term future. All these dissatisfactions are gathered under the rubric of 'alternative agriculture', a term I prefer to avoid for several reasons: for clarity because it covers so wide a spectrum of variations; for understanding because it implies practices not used in conventional modern agriculture; and for greater hope of progress, because it tends to be confrontational whereas it is beneficial for alternative and mainstream agriculturists to exchange views.

Traditional agricultures are still practised in many parts of the world and, with some evolution, are likely to remain so particularly in adverse environments where they are often stable, reliable and subtly adapted. Moreover, as Miguel Altieri[3] argues, there is much to be learned from them about the complex and often unsuspected interactions between people, crops, animals and soils, as he illustrates for the buffalo wallows of Sri Lanka.

If the world population were only three billion, a largely self-sufficient traditional agriculture would be possible, e.g. with two billion hectares under cultivation, half of it in cereals yielding one tonne of grain per hectare. But the population is already twice that and intensification is unavoidable. Moreover, subsistence farmers tend to have large families which help with the farm but result in a growing population which needs ever more arable land. Consequently, even subsistence farming is not sustainable in the long term, and often harmful to the environment in the short term.

A degree of mechanization could reduce the need for family labour, and is acceptable in principle to many alternative agriculturists. Amish farmers in the USA have largely retained the labour-intensive methods and frugal lifestyles of their 16th century forebears but in some regions they now use at least stationary motors. Likewise, while cherishing the diversity of older crops and varieties, alternative agriculturists often use modern varieties because of the wider range and relevance of their genetic resistances to current biotypes of pests and diseases. Beyond this point, however, there is a wide spectrum in the extent to which industrial inputs such as fertilizers, insecticides, fungicides, herbicides and regulants are eschewed. Truly 'Organic' farms avoid all of these, but some 'organic' farms do not.

Apart from lime and some mineral dressings there is a heavy reliance on farmyard manure and on crop rotations with N-fixing legumes to maintain soil fertility and high yields, as there was in the Norfolk four-course rotation. Access to the requisite large amounts of farmyard manure is not always available, and the energy cost just of hauling and spreading manure from feed lots can be half the total energy cost of manufacturing and using nitrogenous fertilizers[159]. The outcome of using the one or the other varies with both crop and conditions. Maize and soybeans respond better to farmyard manure than does wheat. And, for several crops, in poor seasons or at low levels of application, farmyard manure gives higher yields than fertilizers, but the reverse can be the case in good seasons or with high levels of application[121].

The movement away from industrial inputs in agriculture began with the recognition that pesticide residues could be harmful to wildlife, to ecosystems and to people, following the publication of *Silent Spring*. Biological control, the breeding of resistant varieties and, more recently, the development of integrated pest management techniques have made it feasible to reduce or even eliminate the application of pesticides to many crops except where market appearance is important, as with many fruits and vegetables. Natural insecticidal preparations, such as neem, or their derivatives, such as the pyrethroids, are favoured alternatives. Control of weeds remains a problem. Organic farms produce food of a kind desired by some, but even in the USA they constitute only about 1% of farms and less than 0.1% of the agricultural area in Europe[17].

Of much greater significance is the shift in modern agriculture towards more sparing, more efficient and more informed use of industrial inputs. The application of pesticides to cotton crops in the USA is now only a quarter of what it was 20 years ago, and application has fallen, less spectacularly, with other crops and for herbicides as well. Integrated pest management is increasingly used in 'mainstream' agriculture. Environmental concerns combined with the need to reduce costs of production are also leading to more effective use of nitrogenous fertilizers by many farmers, resulting in a substantial drop in fertilizer use in Europe in recent years. The use of crop rotations, to take advantage of biological N-fixation by legumes and to control soil-borne pests and diseases, is also likely to 'ecologize' modern crop production. However, the ecological ideal of a wholly self-sustaining agriculture is unlikely to be possible when there are ten billion people to be fed. There is simply not enough land for the requisite fuel crops, although solar absorbers and wind-powered generators may go some of the way.

So far, the intensification of agriculture has referred primarily to the heavier use of a wider range of inputs. A better form of intensification is now under way, thanks partly to the ecological idealists and partly to economic pressures, namely the more intensive application of our understanding of exactly when and where and how these inputs contribute to greater crop yields. The National Research Council report on *Alternative Agriculture*[145] highlighted the need of farmers for information and technical assistance in developing better crop management practices and skills. It also pointed out that many government policies 'work against environmentally benign practices and the adoption of alternative agricultural systems'.

## 12.10 Dilemmas for agriculturists, young and old

In a book entitled *From Faust to Strangelove* the author, Roslynn Haynes, argues that the image scientists have of themselves is far removed from those held by the public or portrayed by writers of fiction. No doubt this is true also of agricultural scientists, in all their variety, who see themselves as helping to feed and clothe the rapidly growing human population while cherishing the earth, yet are often viewed as destroyers of Nature, wastrels of water, eroders of land and genetic resources, polluters of the environment and hand-maidens of agribusiness.

Wherever the truth lies, I know that many agricultural scientists have agonized on the horns of several dilemmas throughout their working lives. Some of these dilemmas follow:

(1) *The Malthusian dilemma.* Although the views of Malthus on population evolved, the tendency for populations to increase towards the limit of subsistence remained a key element in his argument. Many agriculturists have asked themselves whether, by helping to increase food production, they were simply enlarging the ultimate human population and, therefore, increasing the drain on resources and extending the impact on Nature, rather than responding to greater population pressure by enhancing the food supply, as Boserup envisaged. Malthus' argument that for population pressure to drive us on, it must first have risen towards its limit, continues to haunt us.

(2) *The demographic transition dilemma.* Demographers in the mid–20th century were much concerned with national transitions from high to low birth rates, the second part of their demographic transition, the first being that from high to low death rates. France, the first nation to achieve its fertility decline, did so before the spread of industrialization and urbanization and without any rapid change in agriculture[205]. Perhaps the most important elements of this transition are the widespread perception of reduced human fertility as advantageous, and the availability and acceptance of contraceptive techniques. Increase in agricultural productivity may play a key role in reducing the perceived optimum family size in rural populations in conjunction with the step from a subsistence to a cash economy. For this strategy – i.e. that 'development is the best contraceptive' – to succeed, however, the agricultural advances need to be decisive and quickly and widely adopted, as they were in the Green Revolution. Where they are less decisive, agriculturists remain unsure of their impact on the reduction of birth rates.

(3) *The environmental adversity dilemma.* Agriculturists understandably wish to work in those environments where the hope of progress is higher. Many are aware that a lot of the rural poor live on the less favoured lands in developing countries with poor infrastructure, little access to inputs and advice and, too frequently, degraded land, and they would like to help. But they also recognize how slow and uncertain research for adverse environments can be, and that agricultural

research cannot be expected to solve environmental or social inequities.

(4) *The scale-positive dilemma.* Virtually all agricultural innovations are 'scale-positive' in that they are more quickly and readily available to the larger-scale farmers who have better access to the requisite information, resources and credit. In that sense they support J.B.S. Haldane's assertion that development tends to magnify injustices. Should we therefore *not* try to improve the conditions of the poor and hungry simply because others may benefit even more? Several studies have shown that while larger farmers initially gained more from the Green Revolution, even the smallest farmers eventually gained too, while the urban poor also profited from the greater availability and lower real prices of their staple foods[119]. Several of the international agricultural research centres have tried to devise innovations of specific advantage to the poor farmer, but without real success.

(5) *The sustainability dilemma.* Agriculturists are often torn between their concerns about the need for greater food production now in many developing countries and the need to conserve what is left of Nature and of the resources of agriculture – soil, water, nutrients, genetic variety, etc. – for future generations. Although the proponents of intensive agriculture are frequently accused of having little thought for the future, realistic alternatives for six billion of us now, as well as for ten billion in the future, are limited. The poor, including the many shifting cultivators, often cannot afford less erosive or damaging practices. And the wealthier donor countries which demand more emphasis on sustainability seem to have little interest in supporting the long-term experiments which would highlight emerging problems of sustainability. It is no easy task to balance the claims of providing enough food for the present world population against those of ensuring that the global capacity for food production in the longer term future is not compromised.

(6) *The advancing or maintaining dilemma.* Exploratory research leading to new agronomic procedures, better adaptation or higher yield potential is one key to feeding the ten billion. Just as important, however, is the need to preserve and protect the gains from earlier advances, particularly in the face of continuing evolution of the ability of the myriad pests and diseases of crops to overcome the genetic resistances already bred into them. The name of 'maintenance research' often given to such activities has a ring of ho-humness about it, and not only administrators but also young scientists may underestimate how

important and how challenging it can be. Crop yields might still keep pace with population through improved plant breeding, crop protection and agronomy even if rubisco is not 'improved'.

(7) *The optimist/pessimist dilemma.* Agriculturists concerned with the world food supply are often uncertain how to reply when asked about the prospects, not only because their answer depends on whether needs or wants are being considered (e.g. meat for none, some or all), but also because of the implications of giving either optimistic or pessimistic replies. A pessimistic approach could be regarded as more responsible to our long-term future by discouraging rapid population growth – not that a long line of pessimists has succeeded in achieving that – and as being more likely to provoke governmental action. Pessimism wins the headlines and the book contracts, but has often been shown to be mistaken, e.g. in the case of the Paddock brothers' *Famine–1975!*, or luddite. Moreover, persistent pessimism can breed eventual indifference and distrust of 'expert' opinions. Optimism can mobilize hopes and more positive approaches, but may also generate inaction and complacency. It is an easy target for criticism as 'technological' (i.e. too narrowly based) optimism or as 'pessimism without the facts'. So the young scientist, as well as the reader, must find his/her own way between the unrelieved pessimism of Paul Ehrlich and Lester Brown, the 'conditional' optimism of Daniel Hillel and the unconstrained optimism of Julian Symon. Since authors owe it to their readers to declare their dispositions, I am, like René Dubos, a 'despairing optimist'.

Many other dilemmas face the young agriculturist: whether to work at the system or the process level, by empirical selection or by design, on public knowledge or on intellectual property and, of course, that old intellectual class distinction, pure or applied? For this last there is much to be said for J.D. Bernal's recommendation that we should, like feudal peasants, plough the lord's land for half the time and our own for the remainder. Frits Went's ploughing of his own land led to the herbicides which have resulted in zero tillage on that of the lord.

## 12.11 Food, health, education and work for how many?

Many have been asked, and many have estimated, 'How many people can the earth support?' In his book with that title[44] the demographer

Joel Cohen details 64 of these estimates, which range from one billion to one thousand billion. A quarter of them are less than the present population, and another quarter fall between the present and the estimated population of the earth in 2050.

In part, the range reflects how greatly the answer to the question posed above depends on our specifications, e.g. on what standard of living, what kind of diet, how evenly distributed both within and between societies, at what cost to the environment, for how long into the future? etc. It does not take much reflection on these questions to recognize the range and uncertainty of any answer for just the food component in the title of this final section.

We can be reasonably sure of producing enough food for ten billion by 2050. If we could ensure an even distribution of food with less waste and a greatly reduced use of grain for animal feed, we could almost feed the ten billion now. But, as Millman and Kates[136] put it: 'the history of hunger is embedded in the history of plenty'. Our global dietary inequalities and opportunities are unlikely to change rapidly, and the task of raising crop yields in the developing countries fast enough to match their population growth over the next 50 years, especially given the likely adverse effects of climatic change on crop production at low latitudes, will be difficult. Almost certainly several developing regions will not be able to rely on local production and will have to rely to a greater extent on increases in grain production and export by the developed countries. To feed the ten billion without increasing the arable area will require an average cereal yield for the world as a whole of about 5.0 t $ha^{-1}$, i.e. beyond that yet reached by Europe or North America (Figure 32, p. 205). Although we cannot yet see how this can be done, my guess is that it will be done, but only if agricultural research is strongly supported in both the developed and the developing countries.

Besides the direct effect of an enhanced agriculture on the overall supply of food, Mellor[135] and others have emphasized its 'engine-of-growth' effect on development through the flow-on effects of greater agricultural incomes in rural communities. The importance of these downstream linkages has now been shown in many developing regions, such as in the Punjab[13] and for the Muda irrigation scheme in Malaysia[10], and the poor have often, but not always, benefitted[120].

In this book we have been mostly considering food production at the global level despite Robert Chambers' comment that: 'Concerning food, the biggest error has been to see hunger as a problem of total production'. In fact, as Sen[188] has shown, hunger may not even be a problem of *local* production but, rather, of lack of 'entitlement' to food through lack of work, income, public support or aid, particularly for the inarticulate and very poor. In this context, agricultural development may have as much impact on the poor through its linkages as through its direct effects on food supplies. Of the 800 million undernourished poor in the world today, the majority still live in rural areas where such linkages have their major impact, and many of them live in adverse environments.

The United Nations Hot Springs Conference in 1943 proclaimed: 'The first cause of hunger and malnutrition is poverty.' And the first cause of poverty is lack of work, hence Gandhi's statement: 'To the poor man God dare not appear except in the form of bread and the promise of work.' In turn, the ability to work requires one to be sufficiently fed, healthy and educated to perform adequately. Although the *proportion* of the world population which is malnourished is slowly decreasing, the absolute number remains unacceptably high. Even when food availability increases and hunger decreases, malnutrition may not do so because of poor access to sanitation, clean water, health care and education[80]. Women and children are frequently disadvantaged in these ways, particularly through lack of elementary education, despite the high rates of social return on public investment in rural education.

The many interlocking facets of poverty, of which food supply is only one of several 'basic needs', are now well recognized[120]. Increasing global food production will not solve the problems of poverty and malnutrition, but agricultural development in regions where poverty is predominantly rural will help, as will the continuing reduction in the real price of grain for the urban poor. However, the further raising of crop yields to match further population growth without compromising the ability of future generations to meet their own needs will require all the understanding, inventiveness and interaction of farmers, industrialists, agricultural scientists, educators, environmentalists, health care workers and policy makers that have brought us to where we are, and which have been sampled in this book.

# Epilogue

We have seen that Robert Malthus' innovation-pull and Ester Boserup's population-push views of the relation between agricultural development and population growth are complementary rather than mutually exclusive, as Malthus himself recognized. Although framed in a context of subsistence agriculture, both views are still relevant. The falling food production per head in many parts of Africa highlights Malthusian concerns, while the efforts of agricultural scientists to raise yields in response to increasing population pressure in developing countries exemplify Boserup.

However, in regions which have already passed through their demographic transition, such as Europe and North America, other forces are at work in raising crop productivity. Although Malthus recognized the key role to be played by increase in yield per crop, he could hardly be expected to have foreseen the transforming effects of off-farm industrial inputs of fertilizer, energy and agrichemicals on yield levels and on the commercial impetus to intensification. Nor to foresee other consequences of yields well beyond the farmers' own requirements, which allowed them to look to better education and other prospects for their children, and to a lower birth rate rather than the higher one anticipated by Malthus. One consequence of this transformation is that although Malthus knew of 'no instance where a permanent increase of agriculture has not effected a permanent increase of population', Europe now provides a striking example.

Before his time the majority of people in Europe were poor and chronically hungry, their level of food consumption quite comparable with that of many Asians or South Americans today, as was the extent of their dependence on cereals and root crops. Was poverty the cause of their hunger, or the low productivity of their agriculture? Once again there is no clear answer to the question, but low crop yields and low output per farm worker were surely much to blame. That was soon to change with the development of agricultural science after Davy, Liebig, Lawes and others had launched it at about the time the world population reached one billion.

The world population has long since passed the point where reliance on a self-sufficient agriculture is possible. Reaching three billion was the turning point. Since then the increase in food production has not relied on the further clearing of land for agriculture but, rather, on increases in yield from the conjunction of cheaper nitrogenous fertilizers, dwarf varieties, effective herbicides and irrigation. Since 1960 there has been a very close relation between world population and global average cereal yields. The doubling of world population since then has been possible only because of the agricultural research which has allowed increases in crop yield to keep pace with those in population, so far.

At every stage in the history of agriculture, farmers have had to weigh the requirements for food production against those for the long-term sustainability of agriculture and the environment, whether of irrigation and salination in Mesopotamia, of cultivation and erosion in Greece or the Great Plains of North America, of pest control and pollution, of dwarf cereals and genetic erosion, of irrigation and falling water tables, etc. Trade-offs have been the inevitable companion of growing populations.

Similarly, throughout recorded history changes in climate have caused deterioration or forced the abandonment of agriculture in one region or another – as we have seen for Abu Hureyra, Mesopotamia, Harappa, the Peten of the Maya and elsewhere – and to extensive migrations, e.g. of the Arabs. But now we face a global rather than a regional change in climate, the agricultural consequences of which are difficult to assess, but which will almost certainly fall most heavily on the less developed countries at low latitudes.

Two other limitations on the global capacity to produce more food also begin to loom. One is the sheer scale of the resources – particularly of water for irrigation but also of arable land, energy and nutrients – that will be needed on a sustainable basis to feed twice as many people in the still developing countries, especially given the high level of resource appropriation already reached on a global scale and the need for most food production to be more or less local, or at least within country.

The other is that the genetic yield potentials of our staple crops may be approaching their limits unless their capacity for photosynthesis and growth can be substantially improved. However, there is still considerable scope for higher yields through other genetic modifications and better crop management. J.B.S. Haldane once commented that 'although we know so little, that little is *so* important'. But what we don't yet understand may be our greatest resource for the future, *provided* that research to deepen our understanding is sustained: 'Light now, for Use hereafter' as Thomas Sprat put it over 300 years ago.

Feeding the ten billion can be done, but to do so sustainably in the face of climatic change, equitably in the face of social and regional inequalities, and in time when few seem concerned, remains one of humanity's greatest challenges.

> 'Had we but world enough and time
> ..............
> Though we cannot make our sun
> Stand still, yet we will make him run.'

Andrew Marvell (1621–1678)
'To his coy mistress'

# *References*


1. Agcaoili, M., Rosegrant, M.W. (1995). Global and regional food supply, demand, and trade prospects to 2010. pp. 61–83 *in* Population and Food in the Early Twenty-first Century (*ed.* N. Islam). IFPRI, Washington, D.C.
2. Alexandratos, N. (ed.) (1995). World Agriculture: Towards 2010. FAO, Rome & Wiley, Chichester.
3. Altieri, M.A. (1990). Why study traditional agriculture? pp. 551–564 *in* Agroecology (*eds.* R.C. Carroll, J.H. Vandermeer, P. Rosset). McGraw Hill, New York.
4. Ammerman, A.J., Cavalli-Sforza, L.L. (1971). Measuring the rate of spread of early farming in Europe. *Man* **6**, 674–688.
5. Anderson, R.H. (1936). Grain drills through thirty-nine centuries. *Agricultural History* **10**, 157–205.
6. Andres, L.A., Davis, C.J., Harris, P., Legner, E.F. (1976). Biological control of weeds. pp. 481–499 *in* Theory and Practice of Biological Control (*eds.* C.B. Huffaker, P.S. Messenger). Academic Press, New York.
7. Bainton, J.A. (1993). Considerations for release of herbicide resistant crops. pp. 42–52 *in* Volunteer Crops as Weeds (*eds.* R.J. Froud-Williams, C.M. Knott, P.J.W. Lutman). Assocn. Applied Biologists, Wellesbourne.
8. Baum, W.C. (1986). Partners Against Hunger. The World Bank, Washington D.C.
9. Beadle, G.W. (1977). The origin of *Zea mays*. pp. 615–635 *in* Origins of Agriculture. (*ed.* C.A. Reed). Mouton, The Hague.
10. Bell, C., Hazell, P., Slade, R. (1982). Project Evaluation in Regional Perspective. Johns Hopkins, Baltimore.
11. Bellwood, P. (1991). The Austronesian dispersal and the origin of languages. *Scientific American* **265**(1), 70–75.

12. Bennett, H.H. (1939). Soil Conservation. McGraw Hill, New York.
13. Bhalla, G.S., Chadha, G.K., Kashyap, S.P., Sharma, R.K. (1990). Agricultural growth and structural changes in the Punjab economy: An input-output analysis. *IFPRI Research Report* No. 82.
14. Biraben, J.N. (1979). (Essay on the evolution of numbers of mankind) (in French). *Population* **34**(1) 13–25.
15. Blevins, R.L., Frye, W.W. (1993). Conservation tillage: an ecological approach to soil management. *Advances in Agronomy* **51**, 33–78.
16. Bloch, M. (1931). Les Caractères Originaux de l'Histoire Rurale Française. Oslo.
17. Boeringa, R. (1996) (*ed.*). Alternative methods of agriculture: Description, evaluation and recommendations for research. *Agriculture and Environment* **5**, Special issue.
18. Bogucki, P. (1996). The spread of early farming in Europe. *American Scientist* **84**, 242–253.
19. Bonny, S. (1993). Is agriculture using more and more energy? A French case study. *Agricultural Systems* **43**, 51–66.
20. Boserup, E. (1965). The Conditions of Agricultural Growth. Aldine, Chicago.
21. Bourke, P.M.A. (1964). Emergence of potato blight, 1843–46. *Nature* **203**, 805–8.
22. Bray, F. (1984). Science and Civilization in China. Vol. 6. Biology and Biological Technology. Part II. Agriculture. Cambridge University Press, Cambridge.
23. Brewbaker, J.L. (1979). Diseases of maize in the wet lowland tropics and the collapse of the classic Maya civilization. *Economic Botany* **33**, 101–118.
24. Broecker, W. (1997). Will our ride into the greenhouse future be a smooth one? *Geological Society of America. Today* **7**(5), 1–7.
25. Buringh, P., Dudal, R. (1987). Agricultural land use in space and time. pp. 9–43 *in* Land Transformation in Agriculture (*eds.* M.G. Wolman, F.G.A. Fournier). Wiley, New York.
26. Buringh, P., van Heemst, H.D.J. (1977). An estimation of world food production based on labour-oriented agriculture. Centre for World Food Market Research, Amsterdam. p. 46.
27. Buringh, P., Van Heemst, H.D.J., Staring, G.J. (1979). Potential world food production. pp. 19–88 *in* Model of International Relations in Agriculture (*ed.* H. Linneman). North Holland, Amsterdam.
28. Burkill, I.H. (1951). The rise and decline of the greater yam in the service of man. *Advancement of Science, London* **7**, 443–438.
29. Butler, J.H. (1980). Economic Geography. Spatial and environmental aspects of economic activity. Wiley, New York.

30. Butzer, K.W. (1976). Early Hydraulic Civilization in Egypt. University of Chicago Press, Chicago.
31. Byerlee, D., Morris, M. (1993). Research for marginal environments. Are we under-invested? *Food Policy* **18**, 381–394.
32. Byerlee, D., Moya, P. (1993). Impacts of International Wheat Breeding Research in the Developing World, 1966–1990. CIMMYT, Mexico, D.F.
33. Carpenter, K.J. (1994). Protein and Energy. A study of changing ideas in nutrition. Cambridge University Press, Cambridge.
34. Carson, R. (1962). Silent Spring. Houghton Miflin, New York.
35. Cavalli-Sforza, L.L., Menozzi, P., Piazza, A. (1994). The History and Geography of Human Genes. Princeton University Press, Princeton, N.J.
36. Ceccarelli, S. (1994). Specific adaptation and breeding for marginal conditions. *Euphytica* **77**, 205–219.
37. Chambers, R. (1987). Food and water as if poor people mattered: A professional revolution. pp. 15–21 *in* Water and Water Policy in World Food Supplies (*ed.* W.R. Jordan). Texas A&M, College Station.
38. Chang, J-H. (1981). Corn yield in relation to photoperiod, night temperature, and solar radiation. *Agricultural Meteorology* **24**, 253–262.
39. Chang, K.C. (1986). The Archaeology of Ancient China. Yale University Press, New Haven.
40. Childe, V.G. (1934). New Light on the Most Ancient East. The Oriental Prelude to European Prehistory. Kegan Paul, London.
41. Chorley, G.P.H. (1981). The agricultural revolution in Northern Europe, 1750–1880: Nitrogen, legumes, and crop productivity. *Economic History Review* **34**, 71–93.
42. Cipolla, C.M. (1993). Before the Industrial Revolution: European Society and Economy, 1000–1700. Routledge, New York.
43. Coe, M.D. (1964). The chinampas of Mexico. *Scientific American* **211**, 90–98.
44. Cohen, J.E. (1995). How Many People Can the Earth Support? Norton, New York.
45. Cohen, M.N. (1977). The Food Crisis in Prehistory. Overpopulation and the origins of agriculture. Yale University Press, New Haven.
46. Crosson, P., Anderson, J.R. (1992). Resources and Global Food Prospects. Supply and Demand for Cereals to 2030. World Bank Technical Paper No. 184, Washington D.C.
47. Cure, J.D., Acock, B. (1986). Crop responses to carbon dioxide doubling. *Agricultural and Forest Meteorology* **38**, 127–145.
48. Dale, P.J., Irwin, J.A., Scheffler, J.A. (1993). The experimental and

commercial release of transgenic crop plants. *Plant Breeding* **111**, 1–22.

49. Dalrymple, D. (1986). Development and spread of high-yielding wheat varieties in developing countries. USAID, Washington D.C.
50. DeBach, P. (1974). Biological Control by Natural Enemies. Cambridge University Press, Cambridge.
51. Deevey, E. (1960). The human population. *Scientific American* **203** (3), 194–204.
52. Denbow, J.R., Wilmsen, E.N. (1986). Advent and course of pastoralism in the Kalahari. *Science* **234**, 1509–1515.
53. Denevan, W.M. (1970). Aboriginal drained-field cultivation in the Americas. *Science* **169**, 647–654.
54. Dieffenbach, E.M., Gray, R.B. (1960). The development of the tractor. pp. 25–45 *in* Power to Produce. U.S. Department of Agriculture, Washington D.C.
55. Duvick, D.N. (1996). Plant breeding, an evolutionary concept. *Crop Science* **36**, 539–548.
56. Dyson, T. (1996). Population and Food. Global trends and future prospects. Routledge, London.
57. Ernle, Lord (R.E. Prothero) (1912, 1961). English Farming Past and Present. Heinemann, London.
58. Evans, L.T. (1993). Crop Evolution, Adaptation and Yield. Cambridge University Press, Cambridge.
59. Evans, L.T. (1997). Adapting and improving crops: The endless task. *Philosophical Transactions of the Royal Society* (London) Ser. B. **352**, 901–906.
60. Evenari, M., Shanan, L., Tadmor, N. (1971). The Negev. The challenge of a desert. Harvard University Press, Cambridge Mass.
61. Falkenmark, M. (1994). Landscape as life-support provider: Water-related limitations. pp. 103–116 *in* Population, the Complex Reality (*ed.* F. Graham-Smith). The Royal Society, London.
62. FAO (1987). The Fifth World Food Survey. FAO, Rome.
63. Fischer, G., Heilig, G.K. (1997). Population momentum and the demand on land and water resources. *Philosophical Transactions of the Royal Society* (London) Ser. B. **352**, 869–889.
64. Fischer, G., Frohberg, K., Parry, M.L., Rosenzweig, C. (1995). Climate change and world food supply, demand and trade. pp. 341–382 *in* Climate Change and Agriculture: Analysis of Potential International Impacts. American Society of Agronomy Special Publication No. 50, Madison.
65. Flinn, J.C., Duff, B. (1985). Energy analysis, rice production systems and rice research. IRRI Research Paper Series No. 114, 1–11.

66. Frankel, O.H. (1970). Variation – the essence of life. *Proceedings of the Linnean Society of New South Wales* **95**(2), 158–169.
67. Frankel, O.H. (1985–7). Genetic resources. I-IV *Diversity* **7**, 26–29 (I); **8**, 30–32 (II); **9**, 30–33 (III); **11**, 25–27 (IV).
68. Funderburk, J., Higley, L., Buntin, G.D. (1993). Concepts and directions in arthropod pest management. *Advances in Agronomy* **51**, 125–172.
69. Fussell, G.E. (1959). Low Countries' influence on English farming. *English Historical Review* **74**, 611–622.
70. Gautheret, R.M. (1983). Plant tissue culture: A history. *Botanical Magazine* Tokyo **96**, 393–410.
71. Goldstrom, J.M. (1981). Irish agriculture and the Great Famine. pp. 155–171 *in* Irish Population, Economy and Society (*eds.* J.M. Goldstrom, L.A. Clarkson). Clarendon Press, Oxford.
72. Goodman, D., Sorj, B., Wilkinson, J. (1987). From Farming to Biotechnology. A Theory of Agro-Industrial Development. Blackwell, Oxford.
73. Goodman, M.M. (1988). The history and evolution of maize. *CRC Critical Reviews in Plant Sciences* **7**, 197–220.
74. Gould, F. (1990). Ecological genetics and integrated pest management. pp. 441–456 *in* Agroecology (*eds.* R.C. Carroll, J.H. Vandermeer, P. Rosset). McGraw Hill, New York.
75. Greeley, M. (1982). Farm-level post-harvest food losses: The myth of the soft third option. *Institute of Development Studies, Sussex. Bull.* **13**(3), 51–60.
76. Greenland, D.J. (1997). The Sustainability of Rice Farming. CABI, Wallingford.
77. Gregg, S.A. (1989). Foragers and Farmers. University of Chicago Press, Chicago.
78. Grigg, D.B. (1974). The Agricultural Systems of the World. An evolutionary approach. Cambridge University Press, Cambridge.
79. Grigg, D. (1993). The World Food Problem. Blackwell, Oxford.
80. Haddad, L., Bhattarai, S., Immink, M., Kumar, S., Slack, A. (1995). More than food is needed to achieve good nutrition by 2020. IFPRI 2020 Brief, No. 25.
81. Halmann, M. (1990). Synthetic plant growth regulators. *Advances in Agronomy* **43**, 47–105.
82. Harlan, J.R. (1967). A wild wheat harvest in Turkey. *Archaeology* **20**, 197–201.
83. Harlan, J.R. (1971). Agricultural origins: centers and noncenters. *Science* **174**, 468–474.

84. Harlan, J.R. (1976). Genetic resources in wild relatives of crops. *Crop Science* **16**, 329–333.
85. Harlan, J.R., de Wet, J.M.J. (1972). A simplified classification of cultivated sorghum. *Crop Science* **12**, 172–176.
86. Harlan, J.R., Zohary, D. (1966). Distribution of wild wheats and barley. *Science* **153**, 1074–1080.
87. Harris, D.R. (1972). The origins of agriculture in the tropics. *American Scientist* **60**(2), 180–193.
88. Hawkes, J.G. (1967). The history of the potato. *J. Royal Horticultural Society* **92**, 207–224, 249–262, 288–302, 364–365.
89. Hawkes, J.G., Francisco-Ortega, J. (1993). The early history of the potato in Europe. *Euphytica* **70**, 1–7.
90. Hazell, P.B.R., Ramasamy, C. (1991). The Green Revolution Reconsidered. Johns Hopkins, Baltimore.
91. Healy, M.J.R., Jones, E.L. (1962). Wheat yields in England, 1815–59. *J. Royal Statistical Society* **125**, 574–579.
92. Helms, D. (1990). Conserving the Plains: The Soil Conservation Service in the Great Plains. *Agricultural History* **64**(2): 58–73.
93. Hillel, D. (1991). Out of the Earth. Civilization and the life of the soil. University of California Press, Berkeley.
94. Hillman, G.C. (1975). The plant remains from Tell Abu Hureyra: A preliminary report. *Proceedings of the Prehistoric Society* **41**, 70–73.
95. Hillman, G.C. (1989). Late Palaeolithic plant foods from Wadi Kubbaniya in upper Egypt. pp. 207–239 *in* Foraging and Farming (*eds.* D.R. Harris, G.C. Hillman). Unwin Hyman, London.
96. Hillman, G.C., Colledge, S.M., Harris, D.R. (1989). Plant food economy during the Epipalaeolithic period at Tell Abu Hureyra, Syria: dietary diversity, seasonality and modes of exploitation. pp. 240–268 *in* Foraging and Farming (*eds.* D.R. Harris, G.C. Hillman). Unwin Hyman, London.
97. Ho, P-t (1959). Studies on the Population of China, 1368–1953. Harvard University Press, Cambridge Mass.
98. Ho, P-t. (1975). The Cradle of the East. University of Chicago, Chicago.
99. Hodell, D.A., Curtis, J.H., Brenner, M. (1995). Possible role of climate in the collapse of Classic Maya civilization. *Nature* **375**, 391–394.
100. Hofman, J.P. (1842). Das Chemische Laboratorium der Ludwigs-Universität zu Giessen. G.F. Winter.
101. Holmberg, J., Bass, S., Timberlake, L. (1991). Defending the Future. Earthscan/IIED, London.
102. Hymowitz, T. (1970). On the domestication of the soybean. *Economic Botany* **24**, 408–421.

103. Islam, N. (1995) (*ed.*). Population and Food in the Early Twenty-first Century. IFPRI, Washington D.C.
104. Issar, A.S. (1995). Climatic change and the history of the Middle East. *American Scientist* **83**, 350–355.
105. Iversen, J. (1956). Forest clearance in the Stone Age. *Scientific American* **194**(3), 36–41.
106. Jacobsen, T., Adams, R.M. (1958). Salt and silt in ancient Mesopotamian agriculture. *Science* **128**, 1251–1258.
107. Jarrige, J-F., Meadow, R.H. (1980). The antecedents of civilization in the Indus Valley. *Scientific American* **243**(2), 102–110.
108. Jensen, N.F. (1967). Agrobiology: Specialization or systems analysis. *Science* **157**, 1405–1409.
109. Kenmore, P.E. (1996). Integrated pest management in rice. pp. 76–97 *in* Biotechnology and Integrated Pest Management (*ed.* G. Persley). CABI, Wallingford.
110. Lal, R. (1989). Conservation tillage for sustainable agriculture: tropics versus temperate environments. *Advances in Agronomy* **42**, 86–197.
111. Langer, W. (1976). American foods and Europe's population growth 1750–1850. *Journal of Social History* **8**(2), 51–66.
112. Large, E.C. (1940). The Advance of the Fungi. Jonathan Cape, London.
113. Lathrap, D.W. (1977). Our father the cayman, our mother the gourd: Spinden revisited, or a unitary model for the emergence of agriculture in the New World. pp. 713–751 *in* Origins of Agriculture (*ed.* C.A. Reed). Mouton, The Hague.
114. Le Gros Clark, F., Pirie, N.W. (1951). Four Thousand Million Mouths. Oxford University Press, Oxford.
115. Lee, R. (1979). The !Kung San: Men, Women and Work in a Foraging Society. Cambridge University Press, Cambridge.
116. Lenin Academy of Agricultural Sciences (1969). N.I. Vavilov and Agricultural Science. Moscow.
117. Lerner, J. (1992). Science and agricultural progress: Quantitative evidence from England, 1660–1780. *Agricultural History* **66**(4), 11–27.
118. Li, H.-L. (1983). The domestication of plants in China: ecogeographical considerations. pp. 21–63 *in* The Origins of Chinese Civilization (*ed.* D.N. Keightley). University of California, Berkeley.
119. Lipton, M., Longhurst, R. (1989). New Seeds for Poor People. Johns Hopkins, Baltimore.
120. Lipton, M., Ravallion, M. (1995). Poverty and policy. pp. 2551–2657 *in*

Handbook of Development Economics III B (*eds.* J. Behrman, T.N. Srinivasan). Elsevier, Amsterdam.

121. Lockeretz, W., Shearer, G., Kohl, D.H. (1981). Organic farming in the Corn Belt. *Science* **211**, 540–547.
122. Long, A., Benz, B.F., Donahue, D.J., Jull, A.J.T., Toolin, L.J. (1989). First direct AMS dates on early maize from Tehuacan, Mexico. *Radiocarbon* **31**, 1035–1040.
123. Loomis, R.S. (1978). Ecological dimensions of medieval agrarian systems: An ecologist responds. *Agricultural History* **52**, 478–483.
124. McEvedy, C., Jones, R. (1978). Atlas of World Population History. Penguin, London.
125. McLaren, D.S. (1974). The great protein fiasco. *The Lancet* July 13, 93–96.
126. MacNeish, R.S. (1985). The archaeological record on the problem of the domestication of corn. *Maydica* **30**, 171–178.
127. McRae, H. (1994). The World in 2020: Power, Culture and Prosperity. A vision of the future. Harper Collins, London.
128. Malthus, T.R. (1798/1970). An Essay on the Principle of Population (*ed.* A. Flew). Penguin, London.
129. Malthus, T.R. (1798/1989). An Essay on the Principle of Population. Variorum edition (*ed.* P. James) 2 Vols. Cambridge University Press, Cambridge.
130. Malthus, T.R. (1824). Summary View of the Principle of Population. John Murray, London.
131. Mangelsdorf, P.C. (1965). The evolution of maize. pp. 23–49 *in* Essays on Crop Plant Evolution (*ed.* J. Hutchinson). Cambridge University Press, Cambridge.
132. Meadow, R.H. (1996). The origins and spread of agriculture and pastoralism in northwest South Asia. pp. 390–412 *in* The Origins and Spread of Agriculture and Pastoralism in Eurasia (*ed.* D.R. Harris). UCL Press, London.
133. Mellanby, K. (1992). The DDT Story. British Crop Protection Council, Farnham.
134. Melillo, J.M., Houghton, R.A., Kicklighter, D.W., McGuire, A.D. (1996). Tropical deforestation and the global carbon budget. *Annual Review of Energy & Environment* **21**, 293–310.
135. Mellor, J.W. (1976). The New Economics of Growth. Cornell University Press, Ithaca.
136. Millman, S., Kates, R.W. (1990). Toward understanding hunger. pp. 3–24 *in* Hunger in History. Food Shortage, Poverty and Deprivation (*ed.* L.F. Newman). Blackwell, Oxford.

137. Millman, S., Aronson, S.M., Fruzzetti, L.M., Hollos, M., Okello, R., Whiting, V. (1990). Organization, information, and entitlement in the emerging global food system. pp. 307–330 *in* Hunger in History. Food Shortage, Poverty and Deprivation (*ed.* L.F. Newman). Blackwell, Oxford.
138. Mitchell, D.O., Ingco, M.D. (1995). Global and regional food demand and supply prospects. pp. 49–60 *in* Population and Food in the Early Twenty-first Century (*ed.* N. Islam). IFPRI, Washington D.C.
139. Mokyr, J. (1983). Why Ireland Starved. A quantitative and analytical history of the Irish economy, 1800–1850. Allen & Unwin, London.
140. Molleson, T. (1994). The eloquent bones of Abu Hureyra. *Scientific American* **271**(2), 60–65.
141. Moore, A.M.T. (1979). A Pre-Neolithic farmers' village on the Euphrates. *Scientific American* **241** (2), 50–58.
142. Moore, A.M.T. (1985). The development of Neolithic societies in the Near East. *Advances in World Archaeology* **4**, 1–69.
143. Mujica. E. (1995). Terrace culture and pre-Hispanic traditions. *CIP Circular* **21**(2), 11–18.
144. Müller, P.H. (1949). Dichlorodiphenyl-trichlroethane (DDT) and newer insecticides. Nobel Foundation, Stockholm.
145. National Research Council (1989). Alternative Agriculture. Natl. Academy Press, Washington D.C.
146. Needham, J. (1965). Science and Civilization in China. Vol. 4. Physics and Physical Technology. Part II. Mechanical Engineering. pp. 303–330. Cambridge University Press, Cambridge.
147. Nye, P.H., Greenland, D.J. (1960). The Soil Under Shifting Cultivation. Commonwealth Agricultural Bureau, Farnham Royal.
148. O'Grada, C. (1990). Irish agricultural history: Recent research. *Agricultural History Review* **38** (II), 165–173.
149. Overton, M. (1989). Weather and agricultural change in England, 1660–1739. *Agricultural History* **63**, 77–88.
150. Overton, M. (1990). The critical century? The agrarian history of England and Wales 1750–1850. *Agricultural History Review* **38**, 185–189.
151. Parain, C. (1966). The evolution of agricultural technique. pp. 118–168 *in* Economic History of Europe I. The Agrarian Life of the Middle Ages. Cambridge University Press, Cambridge.
152. Perkins, J.H. (1982). Insects, Experts, and the Insecticide Crisis. Plenum Press, New York.
153. Peterson, G.E. (1967). The discovery and development of 2,4-D. *Agricultural History* **41**, 243–253.
154. Peterson, G.W., Bell, J.C., McSweeney, K., Nielsen, G.A., Robert, P.C.

(1995). Geographic information systems in agronomy. *Advances in Agronomy* **55**, 68–105.

155. Phillips, R.E., Blevins, R.L., Thomas, G.W., Frye, W.W., Phillips, S.H. (1980). No-tillage agriculture. *Science* **208**, 1108–1113.
156. Pimentel, D. (1976). World food crisis: energy and pests. *Bulletin of Entomological Society of America* **22**, 20–26.
157. Pimentel, D., Hurd, L.E., Bellotti, A.C. Forster, M.J., Oka, I.N., Sholes, O.D., Whitman, R.J. (1973). Food production and the energy crisis. *Science* **182**, 443–449.
158. Pimentel, D., Dazhong, W., Giampietro, M. (1990). Technological changes in energy use in U.S. agricultural production. pp. 305–321 *in* Agroecology. Researching the ecological basis for sustainable agriculture (*ed.* S.R. Gliessman). Springer, New York.
159. Pimentel, D., Terhune, E.C. (1977). Energy use in food production. pp. 67–89. *in* Dimensions of World Food Problems (*ed.* E.R. Duncan). Iowa State University, Ames.
160. Plucknett, D.L., Smith, N.J., Williams, J.T., Anishetty, N.M. (1987). Gene Banks and the World's Food. Princeton University Press, Princeton.
161. Postel, S.L., Daily, G.C., Ehrlich, P.R. (1996). Human appropriation of renewable freshwater. *Science* **271**, 785–788.
162. Postgate, J.N. (1984). The problem of yields in Cuneiform texts. *Bulletin of Sumerian Agriculture* **1**, 97–102.
163. Prasad, R., Power, J.F. (1995). Nitrification inhibitors for agriculture, health and the environment. *Advances in Agronomy* **54**, 234–281.
164. President's (USA) Science Advisory Committee (1967). The World Food Problem. White House, Washington D.C.
165. Reed, C.A. (1977). Origins of Agriculture: Discussion and some conclusions. pp. 879–953 *in* Origins of Agriculture (*ed.* C.A. Reed). Mouton, The Hague.
166. Renfrew, C. (1989). The origins of Indo-European languages. *Scientific American* **261** (4), 82–90.
167. Repetto, R. (1986). World Enough and Time. Yale University Press, New Haven.
168. Richey, C.B., Griffith, D.R., Parsons, S.D. (1977). Yields and cultural energy requirements for corn and soybeans with various tillage-planting systems. *Advances in Agronomy* **29**, 141–182.
169. Rindos, D. (1984). The origins of agriculture: an evolutionary perspective. Academic Press, New York.
170. Rogin, L. (1931). The Introduction of Farm Machinery. University of California, Berkeley.
171. Roosevelt, A.C., Housley, R.A., Imazio da Silveira, M., Maranca, S.,

Johnson, R. (1991). Eighth millennium pottery from a prehistoric shell midden in the Brazilian Amazon. *Science* **254**, 1621–1624.

172. Roosevelt, A. et al. (1996). Paleoindian cave dwellers in the Amazon: The peopling of the Americas. *Science* **272**, 373–384.
173. Rosegrant, M.W. (1997). Water resources in the twenty-first century: Challenges and implications for action. Discussion Paper No. 20, IFPRI, Washington D.C.
174. Rosegrant, M.W., Svendsen, M. (1994). Irrigation investment and management policy for Asia in the 1990s: Perspectives for agricultural and irrigation technology policy. pp. 402–424 *in* Agricultural Technology. Policy Issues for the International Community (*ed.* J.R. Anderson). CABI/World Bank, Wallingford.
175. Rosegrant, M.W., Agcaoili-Sombilla, M., Perez, N.D. (1995). Global food projections to 2020: Implications for investment. IFPRI, Washington, D.C.
176. Rosenzweig, C., Parry, M.L. (1994). Potential impact of climate change on world food supply. *Nature* (London) **37**, 133–138.
177. Rossiter, M. (1975). The Emergence of Agricultural Science. Justus von Liebig and the Americans, 1840–1880. Yale University, New Haven.
178. Runnels, C.N. (1995). Environmental degradation in ancient Greece. *Scientific American* **272** (3), 72–75.
179. Russell, E.J. (1966). A History of Agricultural Science in Great Britain, 1620–1954. Allen & Unwin, London.
180. Sage, R.F. (1995). Was low atmospheric $CO_2$ during the Pleistocene a limiting factor for the origin of agriculture? *Global Change Biology* **1**, 93–100.
181. Salaman, R.N. (1949). The History and Social Influence of the Potato. Cambridge University Press, Cambridge.
182. Sanchez, P., Benites, J.R. (1987). Low-input cropping for acid soils of the humid tropics. *Science* **238**, 1521–1527.
183. Sarma, J.S. (1986). Cereal Feed Use in the Third World: Past trends and projections to 2000. IFPRI Research Report No. 57, Washington D.C.
184. Sauer, C.O. (1952). Agricultural Origins and Dispersals. American Geographical Society, New York.
185. Scherr, S.J., Yadav, S. (1996). Land degradation in the developing world: implications for food, agriculture, and the environment to 2020. IFPRI, Washington D.C.
186. Schultz, T.W. (1964). Transforming Traditional Agriculture. Yale University Press, New Haven.
187. Scott, J.M. (1989). Seed coatings and treatments and their effects on plant establishment. *Advances in Agronomy* **42**, 43–83.

188. Sen, A. (1981). Poverty and Famines. An essay on entitlement and deprivation. Clarendon Press, Oxford.
189. Sen, A. (1990). Food entitlement and economic chains. pp. 374–386 *in* Hunger in History. Food Shortage, Poverty and Deprivation (*ed.* L.F. Newman). Blackwell, Oxford.
190. Sharma, M., Brown, L., Qureshi, A., Garcia, M. (1995). An ecoregional perspective on malnutrition. 2020 Brief No. 14. IFPRI, Washington D.C.
191. Simmonds, N.W. (1991). Selection for local adaptation in a plant breeding program. *Theoretical & Applied Genetics* **82**, 363–367.
192. Simmons, I.G. (1987). Transformation of the land in pre-industrial times. pp. 45–77 *in* Land Transformation in Agriculture (*eds.* M.G. Wolman, F.G.A. Fournier). Wiley, New York.
193. Slicher van Bath, B.H. (1969). Eighteenth century agriculture on the continent of Europe: Evolution or revolution. *Agricultural History* **43**, 169–180.
194. Smil, V. (1991). Population growth and nitrogen: an exploration of a critical existential link. *Population and Development Review* **17**, 569–601.
195. Smil, V., Nachman, P., Long, T.V. (1983). Energy Analysis and Agriculture: An Application to U.S. Corn Production. Westview Press, Boulder.
196. Sprague, G.F., Eberhart, S.A. (1977). Corn breeding. pp. 305–362 *in* Corn and Corn Improvement (*ed.* G.F. Sprague). American Society of Agronomy, Madison.
197. Stanhill, G. (1976). Trends and deviations in the yield of the English wheat crop during the last 750 years. *Agro-Ecosystems* **3**, 1–10.
198. Stanhill, G. (1981). The Egyptian agro-ecosystem at the end of the eighteenth century – an analysis based on the 'Description de l'Egypte'. *Agro-Ecosystems* **6**, 305–314.
199. Stanley, D.J., Warne, A.G. (1993). Sea level and initiation of Predynastic culture in the Nile delta. *Nature, London* **363**, 435–438.
200. Steensberg, A. (1976). The husbandry of food production. *Philosophical Transactions of the Royal Society* (London) Ser. B. **275**, 43–54.
201. Steensberg, A. (1980). New Guinea Gardens. A study of husbandry with parallels in prehistoric Europe. Academic Press, New York.
202. Stern, V.M., Smith, R.F., van den Bosch, R., Hagen, K.S. (1959). The integrated control concept. *Hilgardia* **29**(2), 81–101.
203. Stuber, C.W. (1992). Biochemical and molecular markers in plant breeding. *Plant Breeding Reviews* **9**, 37–61.
204. Svendsen, M., Rosegrant, M.W. (1994). Irrigation development in South East Asia beyond 2000: Will the future be like the past? *Water International* **19**(1), 25–35.

205. Teitelbaum, M.S. (1975). Relevance of demographic transition theory for developing countries. *Science* **188**, 420–425.
206. Templeman, W.G. (1955). The uses of plant growth substances. *Annals of Applied Biology* **42**, 162–173.
207. Thacker, J.R.M. (1993/4). Transgenic crop plants and pest control. *Science Progress* **77**(3/4), 207–219.
208. Tribe, D. (1994). Feeding and Greening the World. CABI, Wallingford.
209. Turner, M. (1986) (*ed.*). Malthus and his Time. Macmillan, London.
210. Unger-Hamilton, R. (1989). The Epi-Palaeolithic Southern Levant and the origins of cultivation. *Anthropology* **30**, 88–103.
211. van der Merwe, N.J. (1982). Carbon isotopes, photosynthesis and archaeology. *American Scientist* **70**, 596–606.
212. Vasey, D.E. (1992). An Ecological History of Agriculture: 10,000 BC–AD 10,000. Iowa State University Press, Ames.
213. Vermeij, G.J. (1987). Evolution and Escalation. An Ecological History of Life. Princeton University, Princeton.
214. Vitousek, P.M., Ehrlich, P.R., Ehrlich, A.H., Matson, P.A. (1986). Human appropriation of the products of photosynthesis. *Bioscience* **36**(6), 368–373.
215. Waggoner, P.E. (1994). How much land can ten billion people spare for Nature? Task Force Report No. 121. Council for Agricultural Science and Technology, Ames, Iowa.
216. Waterlow, J.C., Payne, R. (1975). The protein gap. *Nature, London* **258**, 113–117.
217. Wendorf, F., Schild, R., Close, A.E. Donahue, D.J., Jull, A.J.T., Zabel, T.H., Wieckowska, H., Kobusiewicz, M., Issawi, B., El-Hadidi, N. (1984). New radiocarbon dates on the cereals from Wadi Kubbaniya. *Science* **225**, 645–646.
218. Wendorf, F., Close, A.E., Schild, R., Wasylikowa, K., Housley, R.A., Harlan, J.R., Krolik, H. (1992). Saharan exploitation of plants 8,000 years BP. *Nature, London* **359**, 721–724.
219. Wheeler, R.O., Cramer, G.L., Young, K.B., Ospina, E. (1981). The World Livestock Product, Feedstuff and Food Grain System. Winrock International, Morrilton, Ark.
220. Whigham, D.K., Minor, H.C. (1978). Agronomic characteristics and environmental stress. pp. 77–118 *in* Soybean Physiology, Agronomy and Utilization (*ed.* A.G. Norman). Academic Press, New York.
221. White, K.D. (1970). Fallowing, crop rotation and crop yields in Roman times. *Agricultural History* **44**, 281–290.
222. White, L. (1962). Mediaeval Technology and Social Change. Oxford University Press, Oxford.
223. White, P.R. (1956). The Cultivation of Animal and Plant Cells. Ronald Press, New York.

224. Wildman, S.G. (1997). The auxin-A,B enigma: scientific fraud or scientific ineptitude. *Plant Growth Regulation* **22**, 37–68.
225. Wilmsen, E.N. (1989). Land Filled with Flies. A political economy of the Kalahari. University of Chicago, Chicago.
226. World Commission on Environment and Development (1987). Our Common Future. Oxford University Press, Oxford.
227. Wrigley, E.A. (1985). Urban growth and agricultural change: England and the Continent in the early modern period. *Journal of Interdisciplinary History* **15**, 683–728.
228. Yan, W. (1991). China's earliest rice agriculture remains. *Bulletin of Indo-Pacific Prehistory Assocation* **10**, 118–126.

# *Acronyms and abbreviations*

| | |
|---|---|
| B.t. | *Bacillus thuringiensis* |
| $C_3$ | plants with Calvin cycle photosynthesis |
| $C_4$ | plants with the $C_4$ photosynthetic pathway |
| CGIAR | Consultative Group on International Agricultural Research |
| CIAT | Centro Internacional de Agricultura Tropical, Cali, Colombia |
| CIMMYT | Centro International de Mejoramiento de Maiz y Trigo, El Batan, Mexico |
| 2,4-D | 2,4-dichlorophenoxyacetic acid |
| DDT | *p,p′*-dichlorodiphenyltrichloroethane |
| DNA | deoxyribonucleic acid |
| FAO | United Nations Food and Agriculture Organization, Rome |
| GCM | general circulation model |
| ha | hectares |
| IFPRI | International Food Policy Research Institute, Washington D.C. |
| IITA | International Institute of Tropical Agriculture, Ibadan, Nigeria |
| ipm | integrated pest management |
| IRRI | International Rice Research Institute, Los Baños, Philippines |
| J | joule (= 0.24 calories) |
| K | thousand |
| M | million |
| MCPA | 2-methyl-4 chloro-phenoxyacetic acid |
| PAG | United Nations Protein Advisory Group |
| rubisco | ribulose-1,5-bisphosphate carboxylase-oxygenase |
| t | metric tonne |
| 2,4,5-T | trichlorophenoxyacetic acid |
| USDA | United States Department of Agriculture |
| WHO | World Health Organization, Geneva |

# Index